W9-CFW-613

esco press

Awareness of
Fundamental and
Emerging Green
Technologies

green
mechanical
systems

Going **Green** Book Series

FERRIS STATE
UNIVERSITY
COLLEGE OF
TECHNOLOGY

GREEN MECHANICAL
COUNCIL

HVAC
Excellence

Awareness of
Fundamental and
Emerging Green
Technologies

green
mechanical
systems

A joint effort of
HVAC Excellence and Ferris State University

Publisher: ESCO Press, 2007

Contributing Authors:

Michael J. Korcal
    Ferris State University
        Assistant Professor HVACR Department

Joseph R Pacella
    Ferris State University
        Assistant Professor HVACR Department

Dr. John Edward Hohman
    HVAC Excellence

Acknowledgements

Special thanks go to many companies and individuals who helped to fill these pages with text, pictures, and graphics.

Graphic Assistance

AU The Graphic Element – Amy Urick – www.urick.com

Technical Assistance from:

Adams Manufacturing – Jeff Dubasak - www.adamsmanufacturing.com

Armstrong International – Will Grindall, Jeff Zachary and Tracy Clupper – www.armstronginternational.com

Bazzani Associates – Ken Van Dyke – www.bazzani.com

Carrier Corporation – Rich Benkowski - www.corp.carrier.com

Cryocel – Victor J. Ott – www.Cryogel.com

Delphi – Joseph Dunlop – www.delpi.com

Energy Star – www.energystar.gov

Fuel Cells 2000 – Jennifer Gangi – www.fuelcells.org

Geothermal Heat Pump Consortium, Inc. – www.geoexchange.org

Koolfog – Brian Roe – www.coolfog.com

LEDTronics – Jordon Papanier - www.ledtronics.com

Radiant Panel Association - www.energystar.gov

Stewart King, Energy Consultant – www.save-energy.org

U.S. Green Building Council – Andre Poremski - www.usgbc.com/

WMEAC – Rachel Hood and Leslie Lock - www.wmeac.org

Zero Flush – Georg Kueng – www.zeroflush.com

CONTENTS

CONTENTS

Introduction

The word "Green" has become a catchall term for all things environmentally or eco friendly. The word "Green" has entered our daily lives. We read about it in the news, see it on TV and hear about it on the radio.

When we attach the word Green to our building's mechanical systems, Green takes on a more precise role in the environmental world. Green now describes energy efficiency and water conservation. For example, we would consider an electrical generating facility green, if it has extremely low emissions. Yet, if it generated power by burning non-renewable fossil fuels, it would not be considered as Green as power generated from photovoltaic (solar power), or wind turbines.

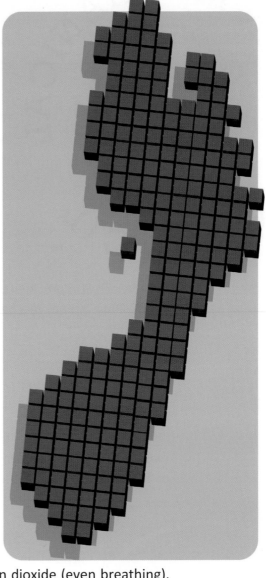

We also play a Green role. Each of us creates a personal carbon footprint. Your carbon footprint is measured by the amount of carbon dioxide that you release into the atmosphere. Carbon dioxide is recognized as a greenhouse gas and is linked to global climatic change. Burning fossil fuels such as coal, oil, and gas causes carbon gases in the forms of carbon monoxide and carbon dioxide to be released into the atmosphere.

Having an understanding of how much carbon will result from various activities, such as the burning of fossil fuels, will allow you to interpret your carbon footprint. Almost every human activity produces carbon dioxide (even breathing). The accumulation of all these activities produces large amounts of carbon. Reducing your personal carbon footprint coupled with the efforts of other people can result in a significant reduction of carbon emissions.

Worldwide, efforts are underway to reduce mankind's carbon footprint. The Kyoto Protocol established mandatory emission limitations for the reduction of greenhouse gas emissions to the signatory nations. As of December 2006, a total of 169 countries and other governmental entities have ratified the agreement.

Green Organizations

In the United States, several organizations have been established to address the Greening of residential, commercial, and industrial buildings.

Green Mechanical Council (GreenMech) is an organization made up of mechanical system industry manufacturers, labor unions, contractor organizations, educators, students, consultants, individual contractors and others.

GreenMech is an international not-for-profit organization formed to focus on energy efficiency for the mechanical systems within our 6 million commercial buildings and 130 million housing units in the United States and Canada. According to GreenMech, "Virtually everyone is a stakeholder in energy efficiency. Each one of us must act now to maximize the efficiency of our existing mechanical systems, and specify high efficiency replacements".

In addition to technical training and certification, GreenMech also provides **Green Awareness** training and certification. GreenMech is establishing green mechanical standards from which a rating system will be launched.

"The benefits of employing Green Mechanical practices are enormous. They include the obvious cost savings, energy independence, new developments in technology, and expanded economic opportunities, while protecting the environment."

"The challenge of the task that lies ahead will require the cooperation of manufacturers, tradesman, educators and the government." www.greenmech.org or Telephone: 1 920 722 4462

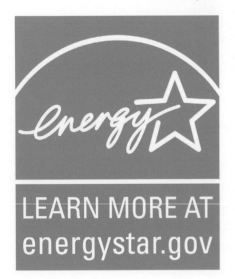

Energy Star is a joint program between the U.S. Environmental Protection Agency and the U.S. Department of Energy designed to save energy and protect the environment. Energy Star was created in 1992 as a voluntary efficiency labeling program for products that consume energy. It is designed as a way to identify energy efficient products being sold.

Computers and computer monitors were some of the first products labeled. Since that time, major appliances, lighting systems, and more have received the label. Homes and businesses are also being labeled as energy efficient by this program.

The U. S. Green Building Council (USGBC) has established LEED® (Leadership in Energy and Environmental Design). This is a Green Building Rating System™ which is a nationally recognized benchmark for the design, construction, and operation of high performance green buildings. LEED establishes standards for the construction of new and the renovation of existing buildings. Buildings built under these standards are considered to be sustainable. The building's systems save water, and energy while maintaining indoor environmental quality.

The Green Building Initiative (GBI)
ww.thegbi.org/greenglobes operates in Canada, USA, UK. GBI's

Green Globe is priced modestly per self-assessment. There is an additional cost for third party verification, which includes a conditional verification at the construction documentation stage, and final verification after a site inspection is conducted.

Leadership in Energy and Environmental Design (LEED), is a Green building rating system based on standards developed by the U.S. Green Building Council (USGBC). A building that meets or exceeds these standards is considered environmentally sustainable. A LEED certified building is a building with low energy usage and low environmental impact.

In addition to LEED certification of buildings, USGBC provides certification exams to individuals testing their knowledge of the LEED rating system. Individuals that pass the exam are recognized as "LEED AP" (Accredited Professional).

FERRIS STATE
UNIVERSITY
COLLEGE OF
TECHNOLOGY

Concepts & Terminology

It was not long ago that terms such as CD, DVD, RAM, hard drive, or software were unknown. In this rapidly changing world, we are constantly adding new concepts and terminology to our lives. The concepts and terms contained in this section are the words and ideas that are used to describe mechanical systems that are Green.

Understanding the world of energy efficiency, water conservation and Green Mechanical Systems requires that you become familiar with some new and some old concepts. Communicating these concepts requires that you speak the language of Green.

The following pages are a list of Green ideas, concepts, words, and terms in use today. This section is not a complete or exhaustive list of terms. It is but a starting place to which we must constantly add information.

Energy Efficiency

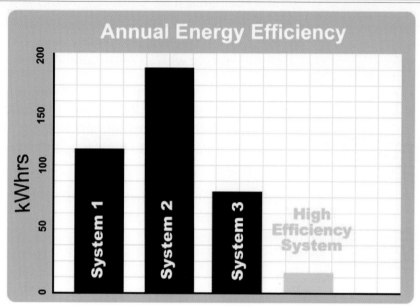

Background

The energy crisis of the 1970s raised the public's awareness of energy efficiency. Energy Efficiency describes how well mechanical systems use energy.

Terms such as Coefficient of Performance (COP), Energy Efficiency Ratio (EER), Heating Seasonal Performance Factor (HSPF) and Seasonal Energy Efficiency Ratio (SEER) describe the energy efficiency ratings of heating and cooling equipment (air conditioners, furnaces, boilers, water heaters, heat pumps, etc.). The efficiency of these various types of equipment is listed on the yellow Energy Efficiency tag required by the Department of Energy (DOE) to be placed on the product. Depending on the type of the system, one of the energy efficiency ratings must be provided.

Why is it Important?

The Energy Efficiency of consumer products should be of interest to everyone. The energy efficiency labels are provided to help consumers make informed equipment purchase choices. The higher the efficiency, the lower the energy cost and energy usage. Product Energy Efficiency ratings help technicians and consumers to effectively compare products.

Terminology

Energy Efficiency – is the relationship of energy consumed to useful work performed. An example of this relationship is the number of watts consumed to the amount of light (lumens) produced.

Annual Fuel Utilization Efficiency

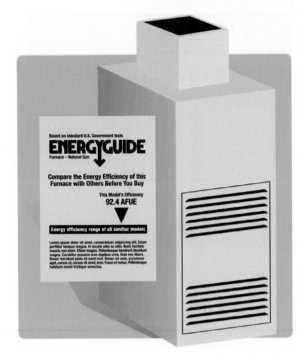

Background

Annual Fuel Utilization Efficiency (AFUE) is a rating of how efficiently a device uses fuel over the heating season. Comfort heating equipment using a fossil fuel is rated and labeled with an AFUE number.

The AFUE numbering system came about as a result of the 1970's energy crisis. Since then, the Federal Trade Commission (FTC) requires all new heating equipment display this number on an easy to read yellow "Energy Guide" sticker.

There is a minimum allowable AFUE rating for various types of systems.

Fossil fueled forced air furnaces	78%
Fossil fueled boilers	80%
Fossil fueled steam boilers	75%

If a heating unit has an 80 AFUE, it means that 80% of the fuel energy is available to heat the occupied space. The remaining 20% leaves the building with the exhaust gas.

Why is it Important?

By comparing the AFUE rating of products, consumers can make informed choices and purchasing decisions. The higher the AFUE number, the greater the energy efficiency of the heating system, meaning less fuel is used to provide the required heat.

Terminology

AFUE - Annual Fuel Utilization Efficiency; a numbering system that relates, in percentage, a heating unit's operating efficiency over the period of the heating season.

Combustion Efficiency - the efficiency of a burner at converting fuel to usable heat energy.

Combustion - the chemical or oxidation process that takes place during burning of any fuel.

Steady State Efficiency - the efficiency of a unit operating at full load.

Thermal efficiency – a measure of the effectiveness of heat transfer in a unit. Thermal efficiency does not take into consideration other losses due to radiation and convection.

Coefficient of Performance

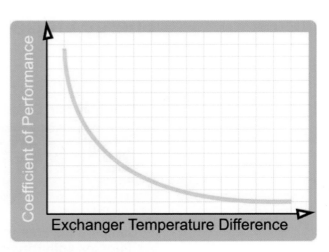

Background

Coefficient of Performance (COP) was established to compare the heating ability of heat pumps. It is the result of a calculation which compares the heating capacity of a heat pump to the amount of electricity required to operate the heat pump.

COP varies from one type of heat pump to another. COP also varies as the outside temperature changes.

The COP will drop for an air-to-air heat pump because it is less efficient at lower outside air temperatures.

The COP of geothermal heat pumps remains higher because ground temperature does not vary as much as air temperature. If a heat pump has a COP of 3, it can move 3 units of heat for every 1 unit of energy it consumes.

Why is it Important?

The COP for some systems is mandated by the Department of Energy (DOE), using recommendations set by the American Society of Heating, Refrigeration, and Air Conditioning Engineers (ASHRAE). COP is used to make comparisons between heat pumps of a particular classification (such as geothermal) and between models. Consumers can use COP ratings to make informed purchase decisions by comparing price and energy efficiency vs. performance.

Terminology

Coefficient of Performance (COP) - the efficiency of a heat pump expressed in terms of heat energy moved for every unit of energy consumed.

Ton - a ton of air conditioning is measured as 12,000 BTU/hour and relates to the cooling effect of 1 ton (2,000 lbs) ice melting in a one day (24 hrs) period.

British Thermal Unit (BTU) - the amount of heat required to change the temperature of one pound of water by one degree Fahrenheit. Melting a pound of ice at 32°F requires 144 BTU.

Watt - a measure of power used by an electrical device. 1,000 watts equals one kW (kilowatt) and one kW equals approximately 1.34 horsepower.

Air to air heat pump – a pump that removes heat from the outdoor air and uses it to heat the occupied space.

Geothermal heat pump – a pump which removes heat from the ground and uses it to heat the occupied space.

Energy Efficiency Ratio

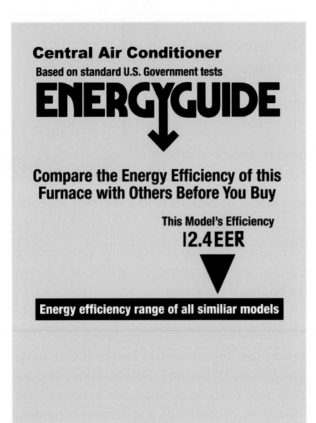

Central Air Conditioner

Based on standard U.S. Government tests

ENERGYGUIDE

Compare the Energy Efficiency of this Furnace with Others Before You Buy

This Model's Efficiency

I2.4EER

Energy efficiency range of all similiar models

Background

Energy Efficiency Ratio (EER) is a measure of how efficiently a cooling system will operate.

The term EER is commonly used when referring to window and unitary air conditioners, as well as heat pumps. An air conditioner is measured for the number of watts of electricity it uses compared to the BTU output.

EER is calculated by dividing the BTUs per hour delivered, by the amount of energy used in watts per hour.

EER ratings are displayed on yellow Energy Efficiency tags as mandated by the Department of Energy (DOE).

Why is it Important?

A higher EER means the system is more efficient. When comparing systems, an EER of 12 would be better than an EER of 8. The higher the EER, the more efficiently power is being converted to useful work. Consumers are able to make comparisons of air conditioning systems by using the EER number.

Terminology

EER - Energy Efficiency Ratio is a ratio calculated by dividing the cooling capacity in BTUs per hour (BTU/H) by the energy input watts. The calculation is based on a given set of design conditions, expressed in BTU/H per watt.

Unitary – a system in which all working parts are enclosed in one housing.

Heating Seasonal Performance Factor

Background

Heating Seasonal Performance Factor (HSPF) is similar to the SEER and Seasonal COP as they all look at the efficiency of the equipment for an entire season.

HSPF takes into account the normal cycling of the systems components such as compressors, fans, etc.

HSPF calculations use the total output of a system and the total electrical power used by a system over an entire heating season.

Why is it Important?

HSPF is one of the energy efficiency rating numbers posted on a product. The rating number can be found on the yellow Energy Efficiency tag required by the Department of Energy (DOE).

The DOE also sets minimum HSPF for heat pumps. ASHRAE recommends HSPF standards to the DOE.

Using the HSPF ratings, consumers can make informed purchase decisions by comparison shopping products and systems. Making wise energy decisions reduces energy waste and lowers energy costs.

Terminology

HSPF – is a measure of the seasonal efficiency of a system (e.g., heat pump) operating in the heating mode, taking into account the variations in temperature that occur within a season and the average number of BTUs of heat delivered for every watt-hour of electricity used.

Seasonal Energy Efficiency Ratio

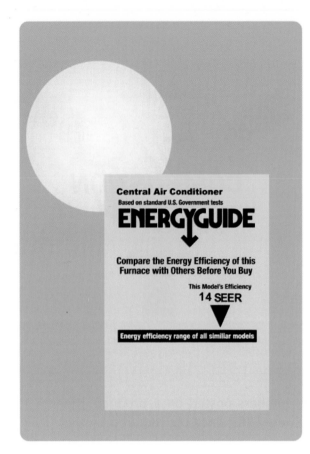

Background

Seasonal Energy Efficiency (SEER) for air conditioning systems is similar to Miles per Gallon (MPG) used for automobiles. Fuel efficiency for automobiles is established for each vehicle based on how it might be driven; city or highway.

In air conditioning systems SEER represents how energy consumed is transformed into useful cooling over a season of use.

Energy Efficiency Ratios (EERs) are used with seasonal data to determine the SEER. Higher SEERs means that a system uses energy more efficiently.

Comparing a system with 10 SEER with one of 13 SEER means that the higher SEER system is 30% more economical to operate.

Why is it Important?

Minimum SEER ratings are established by the Department of Energy. Standards for many residential air conditioning units in the United States began on January 23, 2006. The minimum as of January 23, 2006 is 13 SEER for central air conditioning systems.

Terminology

SEER - Seasonal Energy Efficiency Ratio is a measure of seasonal efficiency of a system compared to the energy required for operation.

Energy Management

Background

Energy Management is the active involvement of a person or system to monitor energy usage patterns. If the term is applied to a system, it generally means that energy management is computer controlled. The computer software is connected to data gathering inputs (sensors).

Energy Management, as applied to a system, can be as simple as a programmable thermostat or as complicated as a system that monitors energy usage patterns, giving operational priority to some energy using devices and shutting others off.

The energy management software is designed to follow programmed requirements and manage energy consumption.

Energy Managers are responsible for purchasing, monitoring, and evaluating energy usage. This person may use energy management systems to obtain information and to control the energy use of other systems in the building or group of buildings.

Why is it Important?

Energy Management, as applied to a person, is important to understand because of the need for individuals with the talent and background to manage energy consumption for large commercial and industrial operations. Energy Managers control a company's energy costs through conservation and energy sourcing.

Terminology

Energy Management – a term used to describe either a system that monitors and controls energy consumption, or a person who monitors an energy management system, purchases energy, manages energy systems, and evaluates energy usage patterns.

Building Information Modeling
Background

Building Information Modeling (BIM) is a complex computer simulation program.

The simulation involves running two distinct programs. The first program is a load calculation program. The second program is a building simulation.

Data is collected about a building or a proposed building prior to operating the program. Data gathered includes:

- building construction & orientation
- windows & doors
- internal loads
- personnel
- utility billing and weather data.

The simulation determines how energy is being used throughout the year and how it will change if an energy conservation measure (ECM) is implemented.

Why is it Important?

BIM supplies critical information about the performance of a building under conditions which simulate that area of the country and regional weather. It provides information about how changes to the building will alter energy use. BIM is also used in the planning and design of new buildings. By understanding how each ECM will reduce energy, the most appropriate and cost effective measures can be made. In this way the largest energy impact can be obtained for the cost.

Terminology

Load Calculation Program – a computer program which uses building construction information and design weather data to determine the heat loss or gain for the building.

Design Weather Data - an outdoor air temperature calculated for both the heating and cooling season, based on historical data.

Building Simulation Program - a computer program which runs a load calculation and a simulation to predict annual energy usage and cost of operation for a building.

Commercial Building Energy Consumption Survey

Background

The Commercial Building Energy Consumption Survey (CBECS) is a national sample survey which collects information on U.S. commercial buildings, their energy related characteristics, energy consumption and expenditures.

A commercial building is considered a building in which half of the floor space is used for a purpose that is not residential, industrial or agricultural in nature. A commercial building is considered to be over 1,000 square feet and used for commercial purposes, such as stores, schools, and office buildings.

Approximately 6,000 businesses are surveyed every few years. CBECS categorizes these commercial buildings by type and use. Under these categories, information on energy consumption for each building type is listed. The first CBECS was completed in 1979 with subsequent surveys conducted every three years. A CBECS was done in the year 2006 and the statistics were made available in 2007.

CBECS information is available to the public and is supplied online by the Energy Information Administration (EIA). There are currently 21 data files which can be downloaded.

Why is it Important?

CBECS are used to compare energy usage to similar commercial buildings. By making comparisons, energy savings estimations of can be calculated.

Terminology

CBECS - Commercial Building Energy Consumption Survey is a national survey of energy use in commercial buildings.

Energy Conservation Measure

Background

An Energy Conservation Measure (ECM) is an action to improve energy efficiency. An ECM addresses the use of excessive energy with a solution that can provide measureable energy savings.

An ECM may be generated as a result of an energy audit by a trained professional, or can be as simple as replacing incandescent bulbs with compact fluorescent lights.

ECMs may address Operation and Maintenance (O&M) issues involving maintenance personnel in procedures that return a system or operation to its original energy efficiency.

Energy Conservation Measures (ECM) typically target:
- Appliances
- Water (energy is needed to pump water)
- Lighting for specific areas.
- Domestic hot water and related controls
- Heating, ventilation and cooling systems and their related controls
- Building envelope: heat loss/gain from roof, walls, floors, doors, windows and operations within the building.
- Fuel switching
- CHP (Combined Heat & Power, or cogeneration)

Why is it Important?

The implementation of ECMs reduces energy consumption and cost.

Planning Energy Conservation Measures provides a blueprint for action. The development of an ECM blueprint aids building owners, facility personnel and energy managers to plan and implement change to the building systems in a logical manner.

Terminology

ECM - Energy Conservation Measure is plan or action for changes to a building or building system that will reduce energy consumption.

Energy Information Administration

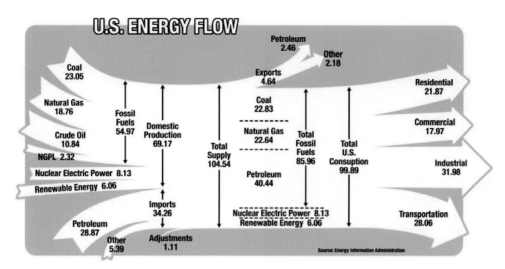

Background

The Energy Information Administration (EIA) is a source for information on energy use in United States. Information is available by type of energy sources, such as petroleum and natural gas.

The EIA provides information on building energy use. An energy forecast is done by the month and by the year. The EIA also estimates the amount of greenhouse gas produced each year from 1949 to the present. All of the information provided by the EIA is available on the web for easy access and download. This is a valuable source of information for everyone.

Why is it Important?

The EIA is the authoritative source for unbiased information concerning energy use. This national agency obtains information directly from consumers, businesses, and energy providers.

Terminology

EIA – Energy Information Administration is under the Department of Energy. It can be found at: www.eia.doe.gov

Forecast – is a prediction based on information gathered from the past to predict the future.

Energy Audit

Background

An energy audit is an in-depth study of a building and its energy consuming systems. Some audits may include the occupants and their interaction with the building and its systems. Energy usage is tracked and documented.

The efficiencies of building systems are determined and the results of an energy audit are used to identify areas of excessive energy consumption.

Why is it Important?

An energy audit is the first step in discovering excessive energy consumption. An audit will show areas of energy inefficiency, and whether these are related to the building, or the building's systems. Understanding where the energy inefficiencies are occurring is the first step to developing Energy Conservation Measures (ECM).

Terminology

Energy Audit – is an assessment of the building and the building systems that use energy.

Energy Consumption and Demand Analysis

Background

An Energy Consumption and Demand Analysis generally follows an energy audit. The analysis is divided into two parts:

The first part looks at the energy use of various types of equipment.

The second part looks at the amount of energy required for systems at specific times during an energy cycle (day, week, or month).

The energy consumed and the amount of energy demanded is compiled into a single report. The consumption analysis also tracks how energy has been purchased.

Why is it Important?

An Energy Consumption and Demand Analysis is an important step in the investigation of energy use and requirements within a building. The Analysis can discover areas where a simple improvement of a system can result in large energy savings. Knowing when there is a large demand for energy allows for efficient planning of systems operations. The analysis also provides information to energy managers, in large commercial and industrial applications, where spot purchasing of energy is desirable.

Terminology

Consumption - The quantity of energy used.

Demand - Total energy required to operate equipment at a given point in time.

Spot purchasing - is the purchasing of energy, based on price, from various sources for as little as six minute time slots.

Heat Load Calculation

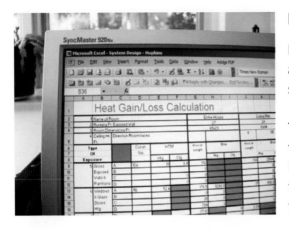

Background

Heat load calculation is used to properly size heating and cooling systems. It includes an evaluation of a structure's ability to resist the transfer of heat.

The calculation uses the R-values of building materials to determine how much resistance to heat transfer is built into the structure, and takes into consideration the outside air temperature and the desired indoor temperature.

Why is it Important?

It is imperative that a "Heat Load Calculation" is performed on all buildings prior to sizing heating and cooling system equipment. Over-sizing or under-sizing heating and cooling equipment can lead to excessive energy consumption, indoor air quality problems, premature equipment failure and inadequate comfort control.

Why not replace a system with the same size system? Changes in the structure (such as new windows, insulation, roof, doors, etc.) can change the system sizing requirements. The addition or maturing of shade trees and additions to the building are also a sizing consideration.

Terminology

BTU - British Thermal Unit is the amount of energy needed to change the temperature of one pound of water one degree Fahrenheit.

Heat loss / gain - Heat that transfers through the building envelope.

R-value – is the thermal resistance to heat flow through a material.

Ghost Loads

Background

Ghost loads are sometimes referred to as "phantom loads" or "stand-by loads" and are points where power is being used without performing useful work. These loads or power consumption points are often transformers that are plugged in, but not powering a device. An example of a ghost load is a cell phone transformer that is plugged in but the cell phone is not attached. Other forms of ghost Loads are systems which appear off, but are on stand-by. Televisions are a good examples of appliances that remain on stand-by. Systems such as these continue to use power. A power strip with a switch can help control ghost loads.

Why is it Important?

Ghost loads are everywhere and waste energy. Knowing the source of ghost Loads and the amount of power they consume is important to the reduction of wasted energy.

Terminology

Ghost Load – is power consumption without producing useful work; also known as a phantom load or stand-by load.

Transformer – is a device that increases or decreases voltage in an AC electrical system, (e.g. 120 volts to 12 volts).

AC – is alternating current. The power from the utility company to a home or business is generally alternating current.

Power – is a reference to the amount of watts being consumed. Watt = Volts X Amps.

Alternative Energy

Renewable and Sustainable Energy

Background

Sustainable energy sources are sources that will not be depleted in a timeframe relevant to the human race, and which therefore contribute to the sustainability of all species.

Renewable energy is a repeatable source of energy whereas the term sustainable is an energy that has no effect or impact on current or future resources. Solar, nuclear, wind, geo-thermal, bio-fuels and water (in the form of rain and waves) are renewable and sustainable forms of energy.

Some fuels used to produce energy are carbon based such as fossil fuels, wood and other bio fuels are Alternative energy

Another example of renewable and sustainable energy is ethanol. Ethanol is made from corn and is a hydrocarbon fuel. Corn is grown with sunlight and is distilled into ethanol (a type alcohol), which can be used to operate internal combustion engines. Solar is 100% renewable, while corn requires energy to produce the alcohol. The net result of alcohol production produces a little more energy than it takes to produce the corn and transform it into alcohol. In addition to sunlight, corn requires fertilizers, soil, and production management to be efficient.

Why is it Important?

Renewable and sustainable forms of energy do not deplete the current sources of hydrocarbon fossil fuels (coal, oil, and gas). Renewable and sustainable energy can reduce our energy dependence on foreign fossil fuel sources, lengthen the amount of time that fossil fuels will be available, and can reduce the amount of carbon released to the atmosphere that is a contributing factor to "green house" effects.

Terminology

Renewable Energy – is a term used for sources of energy that can be used without depleting themselves.

Sustainable energy— is repeatable without effect or impact on current or future resources.

II. Products and Systems

The following section contains some Green mechanical products and systems. This section will give the reader a basic understanding of how these products and systems function. Each product and system has a few words that generally explain the operation or basis for the system. Using the right names and describing the correct operation helps provide clear and accurate communication.

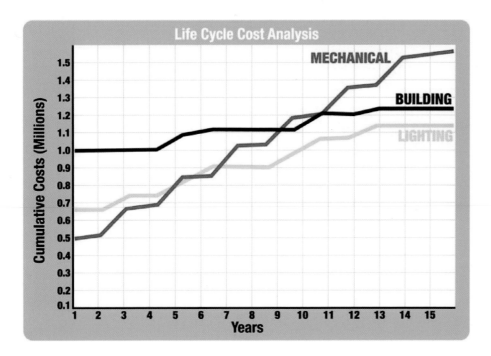

The Life Cycle Assessment or Life Cycle Cost Analysis (LCCA) should to be understood before you specify or purchase Green products and systems.

The initial cost of a product or system, coupled with the expense of installation is frequently 10% or less of the total Life Cycle Cost. The remaining 90% of cost includes the estimated lifetime power / fuel consumption and maintenance costs. LCCA can be used to evaluate the cost of a full range of projects, from an entire site complex to a specific building system component.

LCCA can be used to determine the cost of alternative and energy efficient equipment, sustainable processes, and cost effective solutions to reduce energy use.

Heating, Ventilation and Air Conditioning

The following section on Heating, Ventilation and Air Conditioning (HVAC) identifies high performance low energy consuming products and systems currently available for comfort heating and cooling. Additional concepts and terminology related to HVAC are included.

Control systems for buildings and HVAC systems are not included in this book. Modern control systems are able to modify the operation of standard heating and cooling systems reducing energy usage and increasing human comfort.

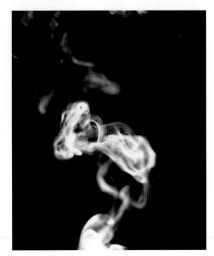

Before addressing HVAC, an overview of **Indoor Air Quality (IAQ)**, the quality of breathable air within a building is necessary. Occupants require air to breathe that is free from odors and hazardous materials. Hazardous materials fall into three major categories including: particulates, biological, and chemical.

Prior to the 1970s and the first major energy crisis, buildings allowed large amounts outside air to infiltrate the structure. This infiltration of outside air helped dilute the indoor air keeping the potential harm to building occupants at a minimum. Attempts to conserve energy by increasing the tightness of buildings, thereby reducing air infiltration, affected indoor air quality. A new concern developed - the indoor air could be hazardous to occupants.

In order to maintain an acceptable indoor air quality, in tighter structures, outside air was brought into buildings. This process (ventilation) is measured in air changes per hour. The American Society of Heating Refrigeration and Air conditioning Engineers (ASHRAE) has established ventilation standards.

Outside air that is allowed to enter a building either by mechanical or passive means is known as ventilation air and is measured in cubic feet per minute (CFM). Whenever air is introduced into a building, the same amount is removed. Uncontrolled air moving into a building is infiltration and when it moves out of a building, uncontrolled air is exfiltrating.

For every action there is an equal and opposite reaction. In our quest for energy efficiency, we must be ever vigilant that our actions do not endanger human health. Airborne particulates, organisms, and vapors affect many people and can cause "Sick Building Syndrome". Air quality within buildings can be controlled to reduce and eliminate potentially hazardous substances. Reducing or eliminating airborne substances increases the healthful benefits of air and reduces medical problems.

Ventilation

Background

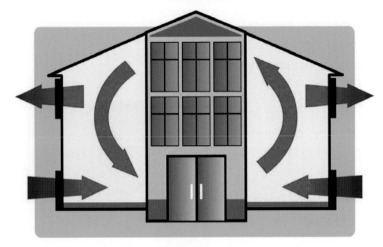

Ventilation is the process of bringing new air into an occupied space.

The amount of ventilation is of concern because this new outside air needs to either be heated or cooled.

Natural ventilation occurs in many older structures due to air infiltration. Newer structures are built tighter than older structures, limiting air infiltration, therefore requiring mechanical ventilation.

Infiltration is uncontrolled whereas ventilation is controlled and the number of air changes is determined based on building use and occupancy.

Why is it Important?

Ventilation is required to increase the amount of fresh air and reduce the amount of carbon dioxide as well as other air-borne contaminants within a space. This air exchange improves the condition of the air quality for the structures occupants.

A large amount of energy may be required to heat and cool the newly introduced fresh air. A balance is needed to reduce energy requirements while maintaining a healthy environment.

Terminology

Ventilation – is controlled fresh air mechanically brought into a structure.

Airborne contamination – includes gasses, vapors, and solids present in the air and are not a natural component of air.

Infiltration — is the uncontrolled entrance of air entering a building.

Comfort Cooling

Comfort cooling systems affect our lifestyle, our longevity, and our social structure. Prior to air conditioning, the extreme heat of summer was often life threatening if not deadly. People spent many hot summer evenings outdoors under the shade of a tree or protection of a porch to escape the heat of their homes.

Comfort cooling systems account for a large share of the energy used during the summer. These systems can strain the electrical power supply. By identifying efficient cooling systems, alternative cooling systems, and building practices that reduce energy load. We will not only save money and energy resources, we will reduce the stress on the power grid.

Mechanical Air Conditioning

Background

Mechanical air conditioning moves heat from a place where it is not wanted to a place where it is less objectionable.

The vast majority of comfort cooling systems in use are air to air vapor compression systems. These systems when properly sized and installed provide comfort cooling and dehumidification.

Why is it Important?

Mechanical vapor compression comfort cooling systems are responsible for the consumption of a large portion of the total summer electrical energy production. Air conditioning equipment owners are becoming more aware of and concerned with the operating efficiency of their systems. With rising electrical energy cost and mandated SEER ratings for new systems, the goal of service personnel must be to minimize energy consumption and optimize performance efficiency for all new and recently repaired systems.

Proper installation and servicing procedures can aid in the reduction of electrical energy consumption resulting in a decrease greenhouse gas emissions.

Terminology

SEER— Seasonal Energy Efficiency Ratio is a measure of seasonal efficiency of a system compared to the energy required for operation.

Saturation Temperature - the temperature at which a liquid will boil. (The boiling temperature of a given liquid varies as pressure changes.)

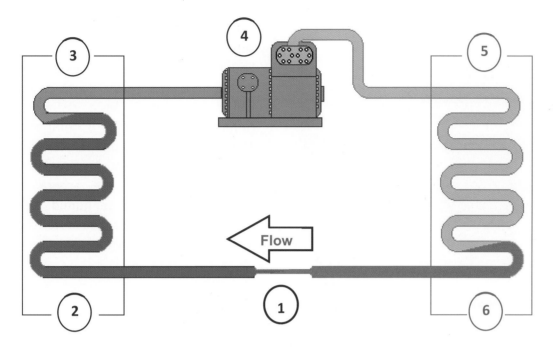

The Mechanical Vapor Compression System

In the vapor / compression refrigeration cycle, liquid refrigerant at a high pressure is delivered to a metering device, (**1**). The metering device causes a reduction in pressure, and therefore a reduction in saturation (boiling) temperature. The refrigerant then travels to the evaporator, (**2**). Heat is absorbed from the air passing over the evaporator and causes the refrigerant in the evaporator to boil from a liquid to a vapor. At the outlet of the evaporator, (**3**), the refrigerant is now a low temperature, low pressure vapor. The refrigerant vapor then travels to the inlet of the compressor, (**4**). The refrigerant vapor is then compressed and moves to the condenser, (**5**). The refrigerant is now a high temperature, high pressure vapor. As the refrigerant transfers its heat to the outside air, the refrigerant condenses to a liquid. At the condenser outlet, (**6**), the refrigerant is a high pressure liquid. The high pressure liquid refrigerant is delivered to the metering device, (**1**), and the sequence begins again.

Evaporative Cooling

Optimal Climate for Evaporative Systems

Background

When water evaporates (changes from a liquid to a vapor), it removes 970 Btu of latent heat energy for every pound of water evaporated. As it changes, the remaining water drops in temperature. The amount of temperature drop depends on the amount of moisture in the air. Drier air will accept more moisture, and wet or humid air will accept far less. Evaporative systems work best in low humidity geographic areas.

Systems

Swamp Cooler

One type of evaporative cooler is referred to as a "Swamp Cooler." Within the Swamp, Cooler water is either sprayed or dripped over the surface of an absorbent material. Air from a fan or blower is passed over or through the material. As the water evaporates in the air stream, the air temperature is reduced. The cool air is directed to the inside the occupied space. Additionally, humidity will rise in the space being cooled. These systems will not work in locations where humidity is already high.

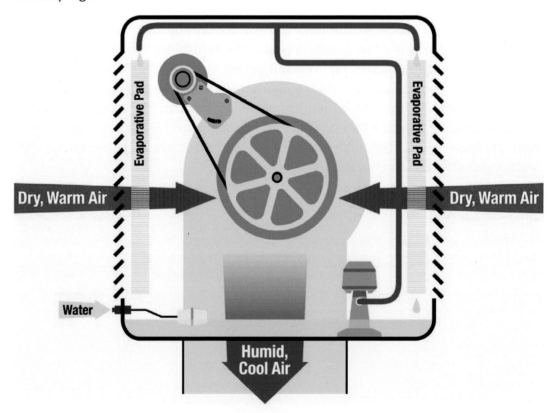

© Koolfog Misting Systems, Inc.

Misting Systems
A fine water mist sprayed into the air can reduce the air temperature and improve comfort. Misters are good for creating cooler areas in very warm industrial locations where attempts at cooling by conventional means would not be economical. Misters are regularly used in outdoor restaurant areas to create a micro-climate area that is much cooler for diners.

Cascade Evaporation

A new development in evaporative cooling ejects the evaporated moisture and heat to the outside. In this way, indoor humidity is not increased. In addition, because this process involves several steps of evaporation (cascade) it tends to drop the temperature in a -more dramatic way than conventional evaporative coolers. This system can approach the cooling capacity of conventional air conditioners when used in areas of low outdoor humidity.

© Delphi, Heat Transfer Div. – Heat & Mass Exchanger

Why is it Important?

Evaporative cooling has been used for centuries. New developments in evaporative cooling conserve water and maximize the evaporative effect. These systems are much lower in energy consumption than standard air conditioning systems. Evaporative cooling combined with standard air conditioning systems can increase efficiency and lower energy consumption.

Terminology

Latent Heat - the heat energy required to change the state of a substance (Example: liquid to vapor) without changing the temperature.

Passive Cooling

Background

Air conditioning is performed by mechanical means and requires energy consumption, whereas passive cooling uses no external energy. There are many passive cooling methods that can reduce the load on a standard air conditioning system, thereby saving money and reducing energy consumption.

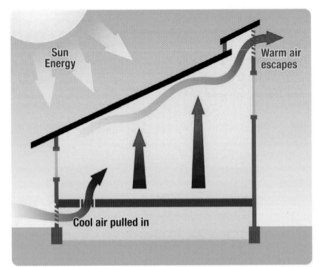

Systems

Open Windows & Solar Chimneys

An open window allows air to blow directly on a person. Air blowing directly on a person's skin will evaporate moisture (sweat) and increase the cooling effect. Opening windows during the night helps to move heat out of the house. Painting the outside wall of a chimney black creates a solar chimney. A solar chimney will absorb the heat of the sun and create an updraft that will aid in removing hot air from a building. There are several other variations of solar chimneys.

Screened Porch

Screened porches do not retain heat and will cool off quickly as the sun sets. The screen helps keep insects out, and allows evening breezes to create a space that can be very comfortable. The screened porch becomes an extension of the living space which does not require traditional air conditioning.

Shade Trees

Summer shade trees that lose their leaves in the fall will allow the winter sun to shine through providing solar heat. In the summer trees reduce the amount of direct solar radiation from entering windows during the day and reduce the energy needed for cooling.

Window Shading

In addition to awnings and interior window shades, there are other window treatments that will reduce the amount of radiant heat entering the living space. Sun screens, in place of regular screens or window tinting can reduce solar radiation through the window, further reducing the energy requirements for cooling.

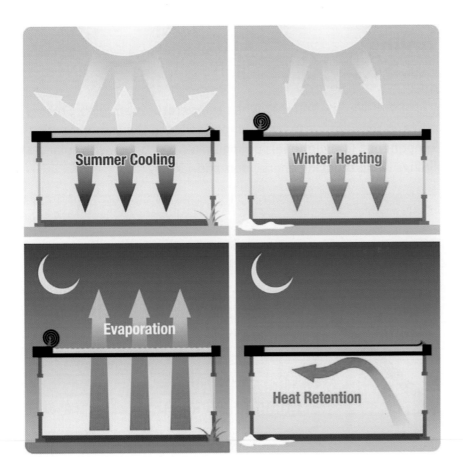

Roof Pond

Roof ponds generally require a flat roof with a load carrying capacity capable of holding a small amount of water. The water is regulated so that as evaporation occurs, the water is replaced. As illustrated a cover is used to regulate heat transfer. Roof ponds can reduce energy requirements for both cooling and heating.

Why is it Important?

Using passive cooling techniques in conjunction with mechanical air conditioning systems can greatly reduce amount of heat entering a structure resulting in a reduction in the amount of energy used.

Terminology

Passive Cooling – is a method of cooling that does not require mechanical energy.

Roof Pond – is a flat roof that contains a small amount of water which evaporates during the day.

Solar chimney - a structure similar in concept to a standard chimney which opens and allow heated air to escape at the top.

Solar Cooling

Background

Solar energy is plentiful in the summer when temperatures are high. Solar systems installed for heat in the winter can be used for cooling in the summer with a few additional components. Conventional cooling systems coupled with solar systems can cool buildings while reducing energy demand.

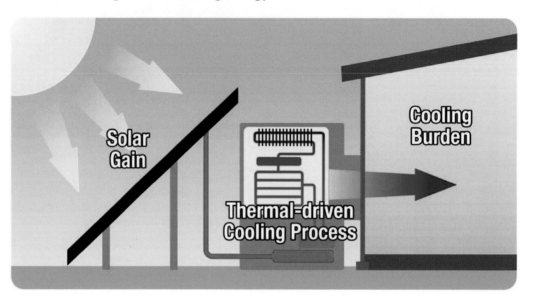

Systems

Absorption

Absorption refrigeration has been around for more than a century. Absorption uses heat to drive the absorption refrigeration cycle. Many RV's (recreation vehicles) have absorption refrigerators which work by burning LP gas or using an electric heating element. These systems are generically known as "thermally driven cooling" systems. Absorption systems for commercial buildings use the hot water or steam that is generated year-round to heat the absorber and create the refrigeration effect. Like boilers, solar collectors produce high temperature hot water. Solar generated hot water is coupled with an absorption system to produce the refrigeration effect needed for comfort cooling.

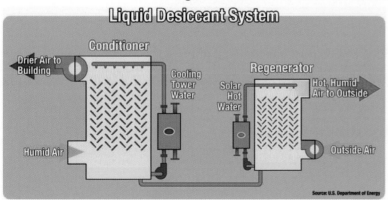

Desiccant

Sometimes referred to as "Sorption" systems, solar desiccant systems are air driers. They use solar heat energy from the collector to drive off moisture in the desiccant container. Driving off moisture regenerates the desiccant which can be

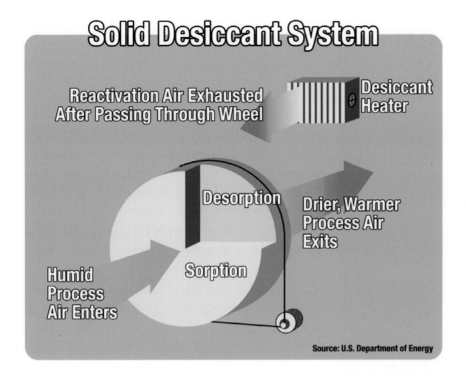

Solid Desiccant System

Reactivation Air Exhausted After Passing Through Wheel

Desiccant Heater

Desorption

Drier, Warmer Process Air Exits

Sorption

Humid Process Air Enters

Source: U.S. Department of Energy

Why is it Important?

Solar energy can be used to power comfort cooling and supplement conventional systems thereby reducing the use of fossil fuels.

Terminology

Desiccant – is a substance which can absorb moisture.

Absorption – also known as a thermally driven cooling system which uses heat energy to drive the cooling process.

Thermal Storage

Background

Thermal energy storage (for example ice storage) is a means of storing energy while it is plentiful to use when peak demand is high. This system takes advantage of reserve refrigeration capability during the night when energy demand is low. At this time, ice is made and stored. When ice forms from water, 144 BTU of latent energy is taken from each pound of water. This is the heat energy it takes to change water to ice at 32 degrees F (or 0 degrees C). During peak demand, the ice is used to create the cooling effect for air conditioning. Some of these systems freeze ice during cold months and then uses the ice to produce cooling during the warmer months of the year. The size of the storage container, the cooling load of the building, and the cost of building the thermal storage system are all taken into consideration in the design phase.

Systems

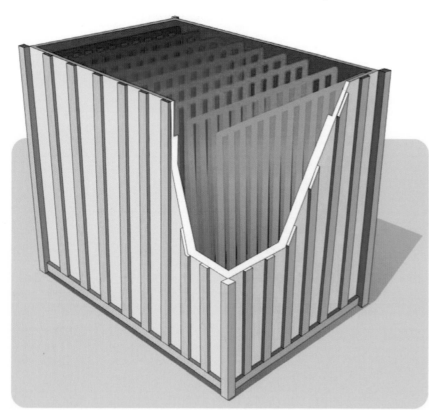

Tank-Type System
Tank-type systems use a tank to hold water and plates or coils to freeze the water. Heat is pulled from the tank and ice forms in the tank and around the plates or coils. When the ice is being used (or melted) the same coils or plates conduct heat energy into the tank.

Cell Type System
Cell-type systems use small, self-contained, plastic balls filled with water. As each ball or cell freezes the. expansion is controlled by the size and shape. Cells are packed into a tank or box where cooling fluid (heat transfer fluid) is passed over each cell. As heat is removed,

each cell freezes. The same fluid is also used carry heat to the tank . As heat is sent back to the tank, the ice in each cell melts absorbing heat from the fluid.

Why is it Important?

By creating a thermal storage system during times when there is a lower energy demand, the stress on the power systems is reduced. Often there is an economic benefit to operating systems during off-peak times to take advantage of reduced electrical costs. If thermal storage could be incorporated into buildings so that ice could be produced during cold weather and used during warm weather, air conditioning operational cost would be reduced.

Terminology

Thermal Storage – is a place where heat can either be removed or stored for later use.

Off-peak – off peak refers to those times during a 24 hour period that electric demand is low.

Latent Energy – also referred to as latent heat; is the amount of energy in BTU that is necessary to change the physical state of a substance from solid to liquid or from liquid to gas.

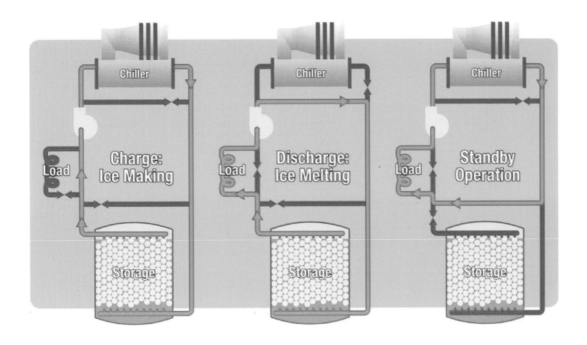

Comfort Heating

Comfort heating systems, like comfort cooling systems, condition our homes and businesses where we work and live. Unlike cooling systems that almost exclusively uses electricity to power the cooling system, heating systems use various fuels. Types of fuels used are gas; liquefied petroleum, natural gas; oil, wood and other biomass, in addition to electricity.

This section will provide information on high performance heating systems and alternative heating systems currently available. By identifying the efficient systems, the alternative heating systems, and the building practices which reduce the cost of heating, we will be able to save money and energy resources.

Thermal Mass and Storage

Background

A thermal mass is anything that will absorb and store large quantities of heat. Large stones and slabs of concrete are good examples. One of the best uses for thermal mass is the absorption of solar heat during the day. At night, as the temperature drops below that of the thermal mass, the thermal mass will transfer its heat back into the space.

The ground is a thermal mass. It may be used to absorb unwanted heat for comfort cooling or transfer heat to homes and businesses during colder weather. Systems that use the ground as a heat sink (thermal mass) are called geothermal systems.

Why is it Important?

The concept and application of thermal mass is important because it is one of the most energy efficient methods of heating or cooling a space. Structures built from materials possessing a high thermal mass absorb heat during the day thereby reducing comfort cooling requirements and release heat at night lowing heating demand.

Terminology

Thermal Mass – is any heavy, dense material (like stone) that will absorb and release heat energy.

Radiation – is energy that moves through air without increasing the temperature of air.

Biomass

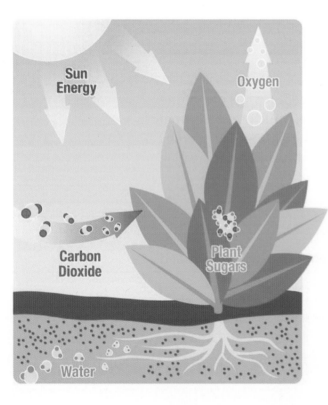

Background

Biomass is any material obtained from anything living or recently living that can be used as fuel. A biomass can be as small as bacteria and as large as a tree. Bacteria and microorganisms produce gasses and reduce materials to other usable products. Trees and other growing things produce products as a result of photosynthesis or the chemical process of growing.

Biomass fuels are carbon based and produce carbon dioxide as a product of combustion.

Systems

There are many biomass types or systems. Below are some of the most popular bio-systems today:

Wood

The most common type of biomass products, wood and waste wood can be used as fuel to produce heat through burning. Burned wood can be gasified to produce "wood gas" and used in much the same way as natural gas. Wood can also be burned in boilers to produce steam. Steam, in turn, can be used to power generators that produce electricity. Heat is the main product of wood biomass. Burning wood in a fireplace or wood burning appliance produces heat.

Solid Waste

Cities and urban areas collect solid waste. Much of this waste is the product of living systems such as wood and food scraps. By collecting and storing these materials properly in landfills, a bio-gas can be produced. Bio-gas is typically called methane and can be used to produce heat. Methane can be used in engines of busses for mass transportation. Methane can also be burned for comfort heating applications.

Bio-fuels

Ethanol is the produced from many types of plants, as well as wood. Ethanol is produced through the fermentation and distillation of plant materials. Currently, ethanol is being added to automobile gasoline. Some engines are designed to run on a mixture of 85% ethanol. Engines of this type are labeled E85. Ethanol can also be burned in comfort heating systems. Another bio-fuel is bio-diesel. Bio-diesel is a fuel that is produced from the waste products of cooking. Many restaurants and food producers use cooking oils to fry foods. These oils are collected, stored and refined to use as a fuel in diesel engines. Bio-diesel can also be used for comfort heating applications.

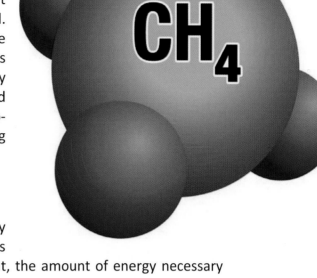

Why is it Important?

Biomass is the product of farming and many other industrial processes. If the biomass produced is used close to the production point, the amount of energy necessary for the production would drop. For example, used cooking oil could be used to power a diesel engine which could produce electricity or supply fuel for comfort heating.

Terminology

Biomass – is material obtained from anything living or previously living.

Photosynthesis – is the process of converting sunlight, water and carbon dioxide into simple sugars for growth in plants.

Microorganism – is a microscopic living thing.

Heating System Efficiency

Background
Depending on the type and size of a comfort system, there may be numerous procedures that can be used to evaluate and optimize a system. Two of the most popular procedures are combustion analysis, and system analysis.

Both, combustion analysis and system analysis will provide information about required to maximize system efficiency.

Systems

Combustion Analysis
Combustion analysis is an essential aspect of boiler efficiency. Combustion analysis should be performed at regular interval and each time a change is made to the combustion system.

Combustion analysis is the measurement of a combustion systems flue gases to determine the completeness of the combustion process. The analysis should be used to return the system to its design specifications.

A service provider may determine the amount of carbon dioxide that is produced at all firing rates of a system. Combustion is the burning of fuel. When a hydrocarbon fuel is burned CO, CO_2 and H_2O are produced. How much CO_2 is produced differs with each type of hydrocarbon fuel and the completeness of the combustion process. The greater the efficiency of the combustion process the greater the amount of CO_2 that is produced.

System Analysis
A system analysis looks at how efficient the system converts fuel to heat. It follows the heating process to determine how efficiently the heat is delivered. System analysis focuses on three aspects of a system:

1) **Generation** – how efficiently the fuel is being converted to heat (combustion analysis);

2) **Distribution** – how efficiently the heat is being distributed.

3) **Return** – the quality and quantity of heating medium being returned to the system.

Why is it Important?

Combustion and system analysis ensures that heating systems are performing safely and at the maximum efficiency according to manufactures design and specification. Maximizing heating system performance is critical to reducing energy consumption, and reducing greenhouse gas emissions.

Terminology

Combustion Analysis – is the process of sampling and analyzing the flue gases produced by a combustion system, to determine how efficiently fuel is being converted to energy.

System Analysis – is the process of reviewing and measuring the entire system operation.

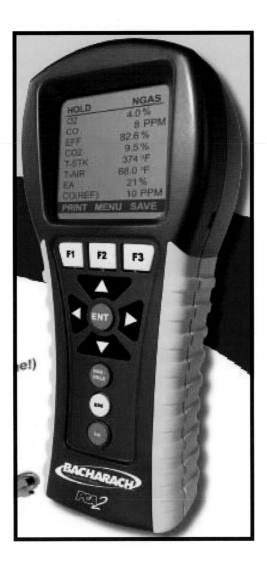

High Efficiency Boiler Systems

Boilers use water as a heating medium rather than air to move the heat from the fuel to the living space. Sometimes called "wet heat", it is understood that in most boiler systems the water is not open to the air. In residential and light commercial boilers, the water is enclosed in the system, where the pressure develops from 12 to 50 pounds per square inch (psi). These 12 to 50 psi boiler systems are called "low pressure hot water" systems. Other commercial and industrial boilers may develop higher pressures. This section of the book highlights several types of high efficiency boiler systems.

Condensing Boilers

Background

A condensing boiler achieves its high efficiencies by increasing the amount of heat exchange surface. The larger heat exchange surface reduces the exhaust gas temperature to the point where water condenses out of the flue gas. Water condensate from the flue gases is collected and sent to a drain. An additional 970 BTU per pound of water condensed provides additional heat for the space. Because condensing water is acidic, heat exchange surfaces, and exhaust gas piping (vent or chimney) needs to be acid resistant. Because exhaust gases are lower in temperature, many venting systems are made from a special plastic. Typical efficiencies are over 90% for condensing boilers.

Systems

Gas Boiler

Condensing gas boilers are the most common. Both natural gas and LP (Liquefied Petroleum) are used as primary fuel sources. These boilers are small and can be mounted on wall surfaces. Many of these are configured in the same way as instantaneous boilers (see Instantaneous Boilers).

Oil Boiler

Condensing oil boilers operate in the same way as condensing gas boilers. Some condensing oil boilers look more like dishwashers and can be placed in living areas of homes. Boilers can also be mounted in cabinets and on walls.

Why is it Important?

As compared to conventional systems operating between 75% and 85%, condensing boilers can operate above 90%.

Terminology

Condensing Boiler – are boilers that extract heat from the exhaust gas until water condenses.

BTU – British Thermal Unit; the amount of heat necessary to raise or lower the temperature of one pound of water one degree Fahrenheit.

Waste-Heat Recovery

Background

Residential and commercial air conditioning and refrigeration equipment remove unwanted heat from a space where it is not wanted and expels it to an area where it is not objectionable, usually outdoors. The expelled heat is wasted (unused) energy. With the addition of a refrigerant to water heat exchanger and circulating pump to an air conditioning or refrigeration system the heat can be captured and used for domestic water heating.

Another method for waste heat recovery in commercial applications is to direct warm air from the refrigeration equipment's condenser, back into the building to assist in comfort heating. This application is ideal for commercial businesses that have food storage equipment.

Why is it important?

Waste-heat recovery can provide much of the domestic water heating needs during the summer months. In commercial applications, waste-heat recovery systems, connected to food storage equipment can provide the water heating needs all year. Based on a 60° rise in temperature, the amount of water heating can be as high as 10 gallons per hour from a 12,000 BTU (1 ton) system, running one hour. The average savings for a family of four can amount to 3500 kWh per year.

Terminology

Condenser - a part of a refrigeration system that rejects the heat from the refrigerant.

Heat exchanger - a device that moves heat energy from one fluid to another (refrigerant to water) while maintaining a complete fluid separation.

Instantaneous Boiler

Background

Instantaneous boilers are similar to instantaneous water heaters. Sometimes the only differences are in the plumbing arrangement and the external controls.

Boilers must be certified by as a boiler by the American Society of Mechanical Engineers (ASME.)

Systems

System Types

Instantaneous boilers are available in similar configurations to that of instantaneous water heaters. System sizing is calculated according load requirements. Some tank-less systems have modulating input heating capabilities. This modulation balances the heating input to the load requirement. The versatile attributes of instantaneous boilers makes them adaptable to low temperature radiant panel heating systems.

Small tank-less

Small tank-less boilers are generally mounted on exterior walls. Larger commercial boilers are floor or rack mounted. Instantaneous boilers are relatively small in comparison to storage type boilers with comparable BTU output.

Why is it Important?

Tank-less boilers are not subject to standby heat losses, and as such are more energy efficient. Water is only heated when there is demand for heat.

Terminology

(See Instantaneous Hot Water for additional terminology)

Radiant Panel Systems

Background

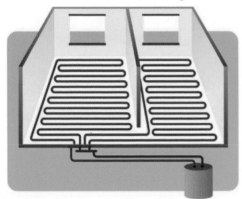

Radiant panel systems can be incorporated into fixtures, floors, walls, and ceilings of structures. Panels typically operate at lower temperatures than conventional systems. Radiant panels provide more even heat than forced air systems. Radiant panels can be operated independently providing excellent zone control. In commercial buildings, radiant systems can significantly reduce the cost of heating by spot heating one location.

Systems

Radiant Floor
Radiant floor heating systems warm floors and transfer heat to the room's objects and occupants. Floors are warm to the touch and operate around 80 to 90 degrees Fahrenheit.

Radiant Ceiling
Radiant ceilings operate at higher temperatures, than floor panels. Radiant energy travels in a straight line to the floor, warming the surfaces of all objects in the room including the occupants.

Radiant Wall
Radiant wall panels have been used effectively where large expanses of wall are available and where retrofitting a floor or ceiling of an existing building is not practical.

Why is it Important?
Radiant panel systems save energy by operating at lower temperatures and by heating objects and occupants instead of the air.

Terminology
Radiant Panel – a panel that is heated and transfers its heat to objects within the space via radiant energy.

Loop – a length of tube placed in various patterns for a given space in which water flows through for heat exchange.

High Efficiency Forced Air Heating Systems

Forced air heating systems heat air by circulating air from the space over a heat exchanger or heating element. Air can be cleaned, humidified, dehumidified, and distributed in by design.

Condensing Furnaces

Background

There are condensing and non-condensing furnaces. Units having AFUE ratings above 90% condense water from the flue gases. The condensing of the flue gas moisture extracts 970 BTU per pound of water condensed. Water condensed from the flue gas must be drained away from the furnace. Condensing furnaces use plastic pipe PVC or ABS approved for use as venting material. PVC or ABS can be used for venting due to the reduced flue gas temperatures in comparisons to that of non-condensing furnaces.

Non-condensing appliances do not condense water from the flue gases and have higher vent temperatures, to prevent condensation in the venting systems.

Systems

Gas

Either natural or Liquefied Petroleum (LP) gas can be used. Gas is burned in an enclosed combustion chamber and flows through a large heat exchanger. As the gasses cool down, water droplets form and collect in the heat exchanger. Heat is extracted and the water is sent to a drain. An exhaust fan pulls flue gasses from the heat exchanger and exhausts them through a plastic vent pipe to the outside. On the other side of the heat exchanger, air from the room is circulated by means of a fan over the high efficiency heat exchanger. Warmed air is then circulated back into the room.

Oil

Condensing oil furnaces work using the same principles as condensing gas furnaces.

Why is it Important?

Condensing furnaces are one of the highest efficiency heating systems. These systems operate at AFUE ratings of 90% or above.

Terminology

Condensation – is the process of changing water vapor to liquid and extracting 970 BTUs per pound of water.

Flue Gas – are gases created when a fuel is burned.

High efficiency furnace with plastic venting

Modulating Furnaces

Background

Modern electronic controls in furnaces have given certain furnaces modulating capability. Modulation means that the furnace heating output will be automatically adjusted to match the heating load as a structure's heat loss increases or decreases. Fuel input and blower speed "modulate" to meet the demand for heat.

Systems

Gas

Modulating gas furnaces make modifications to the amount of fuel burned and blower speed in order to maintain greater efficiency and comfort. The control system anticipates the need for heat based on the feedback it receives through remote sensors.

Why is it Important?

By modulating the heating output to satisfy the load requirement, it reduces the number of start and shut down cycles a furnace goes through. The modulation allows the furnace to operate at a higher efficiency and at a reduced heat output for a longer period, thereby eliminating short cycling. Reducing heat losses encountered during warm up and shut down cycles, reduces energy consumption.

Terminology

Modulating – is the ability to automatically adjust the heating output of an appliance to meet the heating demand of the load.

Heat Loss – is the amount of heat energy that a building loses in the span of an hour.

Solar Heating Systems

Solar heating systems are sustainable energy systems. They operate differently than standard systems and require a different level of routine maintenance.

There is a growing need for technicians to service and maintain these systems.

Solar Air Heating

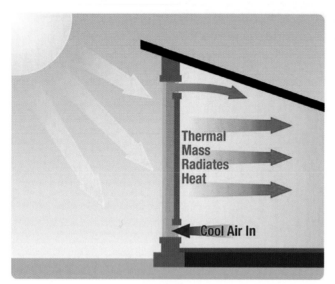

Thermal
Mass
Radiates
Heat

Cool Air In

Background

Background

Solar air heating systems can be either passive or active. Passive systems do not rely on any other energy source to collect and remove heat from the sun. Active systems use blowers and controls to improve their effectiveness. Passive systems rely on components placed in such a way that will allow them to function properly without electricity, while the active systems have components that allow them to be placed in many different configurations. Passive and active systems can be installed in the same building.

Systems

Trombe Wall

A Trombe wall is a passive solar air heating system. The wall is made of a dense material, such as concrete. Glass panels on the outside of the wall allow sunlight to strike the wall. A small space between the glass and the wall allows heated air to rise to an opening at the top of the wall. Openings in the bottom of the wall allow cooler air to enter. The wall's large thermal mass absorbs heat during the day and releases the heat at night as interior temperatures decrease.

Solarium

A solarium is similar to a greenhouse. Most of the walls are glass and the ceiling may be all or partially glass. A solarium is also considered a passive solar heating system.

Sunlight enters the solarium during the day. Solariums need "thermal mass" or dense objects, such as floors and furniture to absorb the heat energy. At night, these dense objects will release energy to heat the space. Some rooms have blinds or shades built in. At night, these shades are drawn to act as an insulator and hold the heat in the solarium.

Active Solar

Active solar air heating systems use externally powered heat circulating components, and come in many different configurations. Some systems have roof-mounted panels. Some panels are on the ground. Thermal storage systems can be simple rock storage bins, and others may exchange heat with water storage systems.

Frequently, the solar air heating panels are installed vertically on south-facing walls with small blowers attached to ducts that transfer heated air directly to the living space. In the winter, sunlight is low on the horizon and strikes the panel. As the year changes to summer and the sunrises higher overhead, overhangs will shade the panel and reduce the amount of heat a panel produces.

Why is it Important?

Solar air heating has the advantage of reducing fossil fuel energy consumption. Windows and large expanses of glass can be used to allow sunlight to enter the space, decreasing the need to use artificial light, and heating sources.

Terminology

Trombe Wall – a massive wall that is heated by the sun and, in turn, heats air by thermal siphon.

Solarium – is a room that is mostly glass.

Thermal Mass – is any heavy, dense material (like stone) that will absorb and release heat energy.

Solar Water Comfort Heating

Background

Solar water systems are designed to heat the interior of buildings and homes. Solar energy is transferred to the water. The solar heated water circulates to a heat exchanger inside the structure. The heat exchanger transfers the heat of the water to the air. The heated water from a solar heating panel could be directly used in a radiant panel heating system. Heat exchangers, pumps, and other components must match the size requirements of the system. Some of these systems also heat domestic water. Fluid in the heating system may be a mixture of water and antifreeze to keep the fluid in the solar panel from freezing in low temperatures.

Systems

Passive

Passive systems use no external source of energy to move solar heated water. Passive solar water space heating systems rely on the sun to heat the water that will flow by thermal siphon effect to the space heat exchanger. The solar collector array must be located below the area being heated. Ground mounted panels are an indication that the solar heating system may be the passive type.

Active

Active systems use pumps to move the solar heated water to obtain solar energy from the collectors. Pumps move the water (or a water glycol mix) through the collector array to a heat exchanger.

Why is it Important?

Solar water heating systems are sustainable systems. Solar energy is continually renewable and provides enough heat energy to completely heat a building or augment standard heating systems. Solar water heating systems aid in the reduction of our carbon footprint.

Terminology

Solar Array – also called a collector array or collector; it is the part of a solar heating system that absorbs heat energy from the sun.

Heat exchanger – any device that transfers heat from one medium to another without mixing the mediums.

Thermal Siphon – is the result of natural convection. Water circulation occurs due to the density differential created by the heating and cooling of water in the piping system.

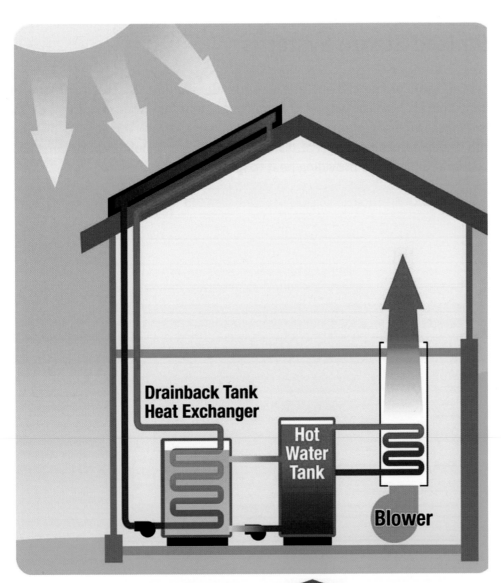

Drainback Tank Heat Exchanger

Hot Water Tank

Blower

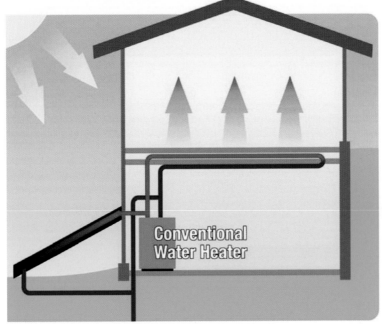

Conventional Water Heater

Optimized Steam Systems

Commercial, light commercial and a few residential systems use steam. Steam is an important and economical way of transporting energy in large quantities or over long distances. Steam generation facilities can be centrally located to distribute steam to multiple buildings at great distances. Some municipalities have centralized steam generation facilities providing heat to local businesses.

Steam Trap management

Background

Condensation within a steam piping system occurs when latent heat energy is removed from the steam and the steam changes back to the liquid state. A steam trap is a device that separates condensate from the steam. Condensate is then returned to the boiler to be turned back into steam.

© Armstrong International, Inc.

Steam traps are designed to trap condensate. Some steam traps fail in the open position allowing steam to pass through the trap. Steam is wasted that could have been used for heat. Some traps fail in the closed position, which will back up condensate. This can cause damage to equipment and the steam system. Facilities that do not manage traps may have an annual failure rate averaging 20 percent. At this rate, there would be 20 failed traps for every 100 high-pressure steam traps resulting in a staggering waste of energy.

Systems

Steam Trap Analysis

Traps are analyzed for functionality by testing using either heat sensing devices or an ultrasonic stethoscope. There are three primary methods of testing steam traps. The techniques include temperature, sound and visual inspection. The simplest and least effective means of testing steam traps is the temperature technique. A more accurate method of testing traps is with sound using an ultrasonic stethoscope. Another accurate approach is the visual method using the test valve that is located on or near the trap and allowing the trap to discharge to the atmosphere.

Steam Trap Management Program

A steam trap management program can be described as a proactive approach to trap maintenance. Traps should be tested on a scheduled basis and test results should be documented. Failed traps must be replaced. An ongoing steam trap management program will reduce energy consumption.

Why is it Important?

Steam trap analysis and management, has the potential to reduce energy consumption dramatically for commercial and industrial heating systems.

Trap management can also prevent over pressurized condensate return lines, premature failure, wear on auxiliary equipment, emergency labor expenses, and unexpected equipment down time.

Terminology

Steam Trap – is a device that automatically separates condensed steam (condensate) from live steam so that the condensate can be returned to the boiler.

Condensate Return

Background

Steam condensate (water) return systems return steam condensate to the boiler where the condensate can be changed back to steam. An Efficient steam system returns steam condensate to the boiler at nearly the same temperature as the steam. Returning the condensate to the boiler while it is hot means less energy is required to turn the condensate back into steam. Most condensate returns are piped to gravity-feed condensate back to the boiler. In other cases, it is necessary for a mechanical pump to return the water to the boiler. There are many condensate return systems that are either not working or not working properly. When a mechanical condensate pumping system fails and condensate cannot get back to the boiler, it leaks out of overflow pipes and is sent to a drain. Scheduled maintenance must be performed to prevent mechanical failure.

Systems

Electric Centrifugal

Electric centrifugal condensate returns rely on the use of a float switch in a receiver tank to turn on the pump motor assemblies.

The most common problems with electric condensate returns are leaky seals and pump cavitations that are directly related to the high temperature of the condensate. Most floor-mounted units are rated to handle a maximum condensate temperature of 200° F. If condensate temperatures rise above the pump's rated temperature the pump may start to cavitate due to the vapor present at the eye of the impeller. This can damage pumps and reduce efficiency requiring additional maintenance and costs.

Non-Electric
Non-electric condensate returns work using a motive force (Steam or Air) to drive condensate out of the pump and into the return line. Mechanical pumps operate with a spring assisted float mechanism, combined with internal motive/vent valves and external check valves.

The most common problem with non-electric (mechanical) condensate return pumps is a lack of motive force.

Why is this Important?
Steam condensate that is not returned to the boiler due to non-functioning steam traps and condensate return systems, increases the volume of makeup water that must be added to the steam system. Make up water added to a system requires more energy than hot condensate to become steam. In addition to the increase of fresh water added to the steam system, additional water treatment chemicals must be added.

Terminology

Condensate Return – a system that is gravity or mechanical (non-electric or electric) that returns condensed water from steam back to the boiler for reheating.

Cavitation - means cavities or bubbles are forming in a liquid, degrading the performance of a pump resulting in a fluctuating flow rate and discharge pressure.

Comfort Heating and Cooling Combination Systems

Geo-thermal, air to air heat pumps and packaged terminal air conditioners (PTACs) provide both comfort heating and cooling. Geo-thermal and air to air heat pumps are configured as either split systems or unitary packages. PTACs are always unitary.

In a split, system the compressor and outdoor coil are located outside of the structure, while the blower motor, controls and indoor coil are located in an air handler inside of the structure. The package or unitary unit has the indoor and outdoor coils, compressor, blower motor and controls housed inside the same cabinet.

These systems use a vapor compression cycle for cooling. Geo-thermal and air to air heat pumps (including heat pump equipped PTACs) reverse the vapor compression cycle for heating, extracting heat from the ground or air to provide primary heat. Where required, electric resistance strips or a gas burning system provides auxiliary heat. PTACs equipped with standard air conditioning provide heat by electric resistance strips or a gas burning system.

These heating / cooling systems have individual SEER, HSPF and/or AFUE ratings for efficiency comparisons and consumer selection.

Geothermal

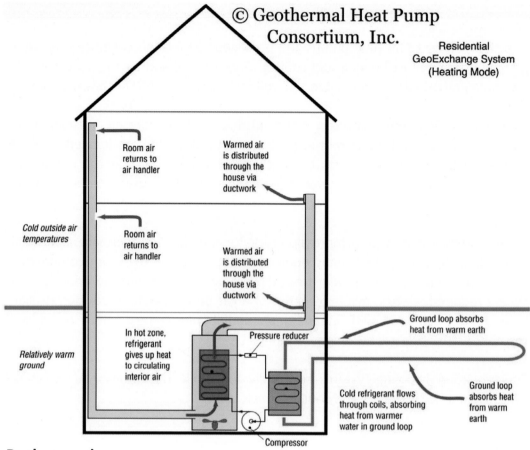

© Geothermal Heat Pump Consortium, Inc.

Residential GeoExchange System (Heating Mode)

Room air returns to air handler

Warmed air is distributed through the house via ductwork

Cold outside air temperatures

Room air returns to air handler

Warmed air is distributed through the house via ductwork

Relatively warm ground

In hot zone, refrigerant gives up heat to circulating interior air

Pressure reducer

Ground loop absorbs heat from warm earth

Cold refrigerant flows through coils, absorbing heat from warmer water in ground loop

Ground loop absorbs heat from warm earth

Compressor

Background

The ground is a thermal mass. It may be used to absorb unwanted heat for comfort cooling or transfer heat to homes and businesses during colder weather. Systems that use the ground as a heat sink (thermal mass) are called geothermal systems. Water in wells, lakes, ponds, etc. may also be used as a thermal mass in geothermal systems.

Systems that use a fluid (refrigerant) to transfer heat to or from the ground or water are called geothermal Heat Pumps. Heat pumps use a reversing valve to control the flow of refrigerant. There are many varying designs of geothermal heat pumps.

Systems

Single Pass

Sometimes called a pump-and-dump, this system takes water from the ground and passes it through the heat exchanger of the system once to extract or absorb heat. The water is then "dumped" to a separate location. An inexpensive source of water is required for this type of system. The only impact on the water is a change in water discharge temperature, warmer in the summer and cooler in the winter.

Horizontal Loop

Water circulates in a loop installed horizontally in the ground. The length of the loop is determined by the soil type, depth, and the system BTU capacity (approximately 400 feet per 12,000BTUs). Water is continually circulated in a closed loop in order to move heat between the heat pump and the ground. The system requires enough open land to install the designed length of horizontal loop.

Vertical Loop

The water circulates in a loop installed vertically in the ground. The layout of this system is similar to a common well. Water continually circulates as in the horizontal loop system. The vertical loop requires less physical space than the horizontal loop (approximately 250 feet of vertical loop per 12,000 BTUs). Several properly spaced wells may be used to meet the needs of the system's capacity.

Direct Expansion

In the Direct Expansion (DX) system, the refrigerant tubing is in direct contact with the ground. If a refrigerant leak occurs, a specially developed heat exchange system prevents ground contamination. Direct Expansion systems have the advantage of not needing a secondary heat exchanger between the ground and the refrigerant. DX systems are more energy efficient than ground source systems.

Why is it Important?

Geothermal units are among the most efficient heating and cooling systems currently available. Compared to a comparable gas fired heating system, the cost of operation is reduced by 66%. The one drawback is the initial cost of installing a geothermal system. The installation cost can be 5 times the cost of traditional heating and cooling systems. These systems require qualified personnel to install and set up. Since these systems use electricity as an energy source for both heating and cooling, they can be coupled with solar panels and be totally renewable and green.

Terminology

Compressor - is the part of the system that pumps refrigerant.

Condenser - is a heat exchanger that rejects heat; identified as being hot during operation.

Metering Device - is a device, which restricts the flow of refrigerant to a determined amount.

Evaporator - is a heat exchanger that absorbs heat; identified as being cold during operation.

Reversing Valve - is a device, which reverses refrigerant flow through the indoor and outdoor coils for cooling or heating mode of operation.

Packaged Terminal Air Conditioners (PTAC) and Air To Air Heat Pumps

Background

Air to air heat pump systems are similar to geothermal systems in principle. Air to air systems use the air as their thermal mass.

PTACs use a vapor compression system to move heat from a place where it is not wanted to a place where it is less objectionable and generally provide heat from electric resistance strips or a gas burning system.

Systems

Cooling (PTAC and Air to air Heat Pump)

Refrigerant in the system evaporates within the indoor coil when it absorbs heat from the return air passing over the coil. Once heat is extracted from the air and the air enters the occupied space it is called the supply air. The refrigerant having absorbed heat then travels to the outdoor coil were it gives up its heat to the outdoor air.

Heating (Air to air Heat Pump)

A heat pump is very similar to an air conditioner. The difference is the heating mode. A heat pump has a reversing valve, which reverses the refrigerant flow in the coils and refrigerant lines. Refrigerant in the system evaporates in the outdoor coil by absorbing heat from air passing over the coil. The refrigerant then travels to the indoor coil were it gives up its heat to the indoor air.

In order to provide proper dehumidification during the cooling season, heat pumps are sized based on cooling requirements. For this reason, the capacity of heat pumps may not be sufficient to satisfy the entire heating requirement. Where required, electric resistance strips or a gas burning system provides auxiliary heat.

Heating (PTAC)

Some PTACs may contain a heat pump and would operate the same as the system above. PTACs that contain a standard air conditioning vapor compression system require electric resistance strips or a gas burning system to provide heat.

Why is it Important?

Packaged Terminal Air Conditioners provide total zone control in residential commercial applications such as hotels, motels and room additions. This system of total zone control reduces energy consumption by providing comfort heating and cooling to occupied areas on demand.

Air to air Heat Pumps are up to three times more efficient than electric resistance heat. At any outdoor temperature, the coefficient of performance (COP) for electric heat is always 1 to 1. At an outdoor temperature of 47°F most air to air heat pumps have a COP of greater than 3 to 1 and at 10°F heat pumps have around a 1 to 1 COP. Heat pumps when properly installed and sized can reduce energy consumption and our carbon footprint.

In some areas where fossil fuels are not readily available, local code requires that homes be heated by a heat pump instead of electric resistance heat.

Terminology

Refrigerant – a fluid that absorbs heat when it evaporates (changes from a liquid to a vapor) and releases heat when it condensers (changes from a vapor to a liquid).

Package Terminal Air Conditioner (PTAC) - is an air conditioning system in which all components are in a single cabinet.

Electrical

Electricity is commonly produced by burning fossil fuels, which is expensive, non-sustainable, and non-renewable, emits greenhouse gases and adds pollutants to the atmosphere. Various alternative methods of electrical generation are also available such as, solar, wind, tidal, hydroelectric, and nuclear, etc. This ever-growing demand for electricity and the onset of global warming brought on by the continued burning of fossil fuels, has made it necessary to better utilize alternative and sustainable energy production technologies. In addition to producing electricity, using clean efficient and sustainable technologies we must continue to develop new technologies.

In addition to electric generation, it is important to specify and use energy efficient appliances, lighting, entertainment equipment, etc.

With existing clean and new technologies, we can reduce electrical demand, energy consumption, and our carbon footprint.

Electrical Power

Background

The cost for electricity is based on Watts (Power) consumed over a period of time. If a 60 watt light bulb is energized for 5 hours, the power consumption is 300 watt-hours. The cost of energy consumption is billed in kilowatt hours. It takes 1000 watts to equal 1 kilowatt.

There is a mathematical relationship between Voltage, Amperage and Watts.
Voltage X Amperage = Watts (Power)

Why is it important

How well a device converts electricity into useful work determines its efficiency. There are many appliances, electronics and electrical devices that operate with higher efficiency's than similar products available. Manufacturers may have multiple products to do the same job. The consumer should compare the products actual energy consumption to determine which product is most efficient.

Terminology

Voltage — is electrical pressure or force.

Amperage — is the amount of electrical current flowing.

Watt — unit of electrical power measurement.

Kilowatt hour — 1000 watts of electrical power used for 1hour.

Fuel Cells

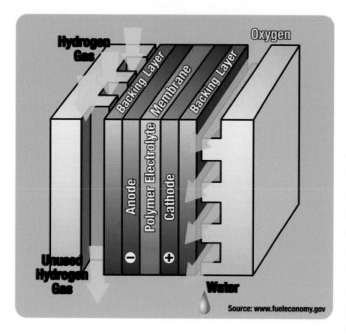

Source: www.fueleconomy.gov

Background

Fuel cells were first developed in1839. NASA is one of the largest users of fuel cells. NASA uses fuel cells to provide electrical power in their spacecraft.

Several different types of fuel cells have been developed. The Proton Exchange Membrane (PEM) is the most popular in use. In the PEM, hydrogen gas and oxygen are passed through a membrane. Hydrogen atoms enter a fuel cell at the anode (positive plate) where a chemical reaction strips them of their electrons. Oxygen from the air is introduced to a cathode (negative plate). The electrons are now available to provide electrical power. When the hydrogen and oxygen combine, water forms as a byproduct.

Systems

Fuel Cells

The most efficient fuel cells use compressed hydrogen and oxygen. Other fuel cell technologies use methane and other hydrocarbons instead of refined hydrogen and air as the source oxygen.

As the hydrogen and oxygen are combined in, a fuel cells heat is produced. The heat generation, from the fuel cell may be used to heat spaces and / or water.

The amount of energy produced by a fuel cell can be regulated to match the energy demand. Matching the energy demand means that only the energy that is needed is produced.

Fuel Cell Stacks

Voltage for a single fuel cell is typically around 1.16 volts. To achieve higher voltage, fuel cells need to be stacked. Fuel cell stacks can be stacked to produce any voltage value. Fuel cells are currently in development that are small enough to be used with cell phones and others that are large enough to be used in commercial operations.

Hydrogen Reformation (reforming)

To produce hydrogen from other fuel sources a process called reformation is used. Hydrogen reformation is a process that converts hydrocarbon fuels (fossil fuels, bio-fuels) into hydrogen. A few systems have fuel cells combined with hydrogen reformers.

Why is it Important?

Traditionally electrical power is distributed from the power generation plant to businesses and homes across a grid of wires. Fuel cells are independent power stations that are not connected to the grid. Fuel cells convert energy into electricity at the point of use thereby eliminating transmission losses. Unlike fossil fuel burning power plants, which produce greenhouse gases, the only byproducts of a fuel cell are water and heat.

Terminology

Fuel Cell - is a battery-like device that produces electricity using oxygen and hydrogen.

Fuel Cell Stack - a number of individual fuel cells connected to produce higher voltage values.

PEM (Polymer Electrolyte Membrane) - a structural component of a particular type of fuel cell; also known as a proton-exchange system.

Hydrogen Reformation (reforming) – is the process of extracting hydrogen from hydrocarbon fuels.

Transmission loss - a loss of electrical power caused by resistance in the grids transmission lines.

Photovoltaic

Background
Photovoltaic (PV) is a term that refers to the process of converting sunlight directly into electricity. Photovoltaic cells also called "solar cells" are the smallest component in a photovoltaic module. When modules are interconnected, they form a photovoltaic array or solar panel. The number of cells connected together determines the amount of electrical power (wattage) that can be obtained from the panel. Solar panels produce DC (Direct Current) electricity. Power obtained from the panel can be stored in batteries or inverted to AC (Alternating Current) electricity.

Systems

Solar Panels
Solar panels can be purchased with specific voltages to suit an application. They are also sized to match the power output requirements. There are many types of cells in production and others in development. Cells can be formed inside of ridged glass or flexible plastics.

Battery Charging Systems
While some systems produce electrical power for direct use, many photovoltaic applications are designed to charge a battery during the day so that the stored power can be used at night.

Portable Power Systems
Photovoltaic (PV) arrays can be brought to a location where the power grid is not available. Panels can be set up to capture the sun and provide power at the location for as long as necessary.

Why is it Important?
Photovoltaic (PV) arrays reduce our dependence on fossil fuels, and oil imports, while protecting the environment. From residential driveway lighting to the primary electric power source for a large high-rise building Photovoltaic (PV) arrays are providing the required power.

Terminology
Photovoltaic (PV) – also called solar cell or array, the word refers to converting "light into electricity."

Grid – refers to power distribution from power plants through a network of wires.

Wattage – is a measurement of electric power.

Photovoltaic (PV) arrays
(Solar Panel)

Parking ticket meter

Wind Turbines

Background
Wind turbines are different than windmills. Windmills pump water or do work directly from the wind. Wind turbines produce electricity.

Wind is caused by the interaction of sunlight on landmasses that create high and low weather areas. Increasing the wind speed one mile per hour, increases the amount of power derived from a wind turbine three fold.

Wind farms have many wind turbines and may provide power to the grid. With the proper amount of wind, a single wind generator can produce the electricity for a single house or small business.

Wind turbines can produce both alternating current (AC) and direct current (DC). DC wind turbines are generally used to store electricity in batteries whereas AC wind turbines are used for immediate application.

The use of a wind turbine for a specific application requires a match between the amount of electrical power required and the size of the turbine. Wind turbines are sized by the kilowatts they generate. Because wind turbines only work when the wind blows, the size of the turbine is determined by using the average wind speed. To obtain the average wind speed, a wind evaluation or study needs to be done at the location.

Systems
Wind Farm
A wind farm is a business that installs, maintains and operates wind turbines for the express purpose of generating electrical power. These farms can be owned by traditional power companies or be independent power producers dealing in wind energy. Wind farms are generally established in areas where there are few trees, buildings, and other obstructions. Wind farms have been producing power for more than 10 years. Cities have also installed one or more turbines to augment power.

Single Turbine Wind Systems
Single turbine wind systems are usually designed to power a single home or business. Single turbine systems are chosen for a home or business based on

amount of electricity required.

Why is it Important?

As a sustainable source of energy, the harnessing of wind power for the production of electricity reduces our carbon footprint while maintaining our lifestyle. We should all welcome the look of wind turbines on the horizon.

Terminology

Windmill – is a device that produces mechanical power for direct use.

Wind turbine – is a device that produces electricity from wind.

Wind Farm – is more than one turbine used to harvest energy from the wind to produce electricity.

Motor Efficiency

Background

Motors can consume large quantities of electrical energy. Two common HVAC uses for electrical motors are to move air and to pump fluids. Air and water movement requirements within a building are never constant. As a result of these changing requirements, there must be changes in the flow of fluids. Controlling the motor speeds provide a solution to the changing needs of the structure and will result in energy savings.

Systems

Electronic Commutated Motor (ECM)

Compared to other single-phase motors, ECM motors have the highest efficiency ratings and are used in many residential air conditioning systems. This motor is a Direct Current (DC) motor, but it has no brushes. It is controlled by a built-in variable frequency drive. The frequency output of the drive allows for variable speeds of the motor, which can be adjusted for the heating and cooling demand, as require. When this motor is used within an air handler, the dip switches on the control board can be set to meet the cooling or heating CFM requirements. When used with the newer electronic thermostats, which monitor the indoor relative humidity as well as the temperature, the blower speed will vary to help control the humidity. The cubic feet per minute (CFM) of air can also be varied for stages of heating and cooling. The motor also saves energy by running at 50% of the cooling speed when the thermostat fan switch is set to the continuous position.

Variable Frequency Drives (VFDs)

Variable Frequency Drives are a common piece of electrical equipment found in most high efficiency commercial systems today. VFDs operate motors more efficiently by increasing or decreasing the speed of the motor by changing the voltage and frequency.

Variable Speed Drives (VSD)

Variable speed drives can be used on DC (direct current), single phase and 3 phase AC motors. They differ from variable frequency drives in switching action. The principle mode of operation is pulse with modulation. The amount of time the switches are opened and closed is used to vary motor speed. The controls can range in voltages from 110 to 10Kv.

Why is it Important?

It is important to understand the relationship between the fluid flow (air or water) and the power needed to move it. By lowering the flow of a fluid within the building to match its present need, the power consumption of the motor will decrease and result in significant energy savings. ECM, VSD and VSD controlled motors are available energy saving alternatives.

Terminology

ECM.— Electronic Commutated Motor is the highest efficiency single phase motor used today in residential HVAC equipment. It is a brushless DC motor operated with an attached or built-in variable frequency drive.

VFD – Variable Frequency Drive (sometimes referred to as Freq-Drive) is an electronic device that modifies sine wave of the voltage being sent to a motor to vary motor speed.

VSD — Variable speed drives controls the amount of time the switches are opened and closed is to vary motor speed.

Lighting

Due to improvements in technology and materials, higher efficiency lighting sources are readily available. Fluorescent and LED lights are being improved and used because of their low electrical consumption.

Historically, most lighting has been purchased based on wattage of the bulb. A 100-watt incandescent bulb produces more lumens (light) than a 60-watt incandescent bulb.

Modern bulbs are rated in lumens. One lumen is the amount of light on 1 square foot of surface created by a candle at a distance of 1 foot. A 60-watt incandescent bulb produces approximately 850 lumens of light. When comparing different types of bulbs, the higher the lumens the greater the light output.

Fluorescent

Background
Fluorescent lights work by using high voltage and high frequency electricity to create ultraviolet light that reacts with a phosphor coating on the inside of a tube. The phosphorus coating can be made of a mixture of many different materials to give various colors of light.

Systems
Fluorescent Tube
Standard fluorescent tubes are illuminated by using high voltage transformers called ballasts. Modern ballasts do not contain wound transformers. Instead, ballasts are made with electronic components that reduce the amount of electricity consumed. Some electronic ballasts (dimming ballasts) allow fluorescent tubes to be operated at lower power levels, which reduce the amount of light when total illumination is not necessary.

Compact Fluorescent
Compact fluorescent lights can be thought of as fluorescent tubes that have been bent and curled to fit the same space as a standard incandescent (filament-type) bulb. These operate in the same way as the fluorescent tube. The ballast for the bulb is located in the base, above the screw threads and below the bulb.

Why is it Important?
Fluorescent lights use approximately 1/4 of the electricity that standard incandescent lights use. They also have a much longer life expectancy than standard incandescent bulbs. Therefore, fewer bulbs are replaced over a greater span of time and the amount of electricity consumed is reduced.

Example;

An incandescent bulb uses 60 watts of power to produce approximately 850 lumens of light and has a life expectancy of 1500 to 2000 hours. This type of bulb will produce around 14 lumens of light per watt of power.

A compact fluorescent light uses 13 watts of power to produce approximately 900 lumens of light and has a life expectancy of up to 10,000 hours. This type of bulb will produce around 69 lumens of light per watt of power.

Terminology

Fluorescent Light – is a tube that illuminates a phosphor coating on the inside of the tube, using high voltage and high frequency electricity.

Compact fluorescent — is a lamp and ballast assembly designed to replace a standard incandescent bulb.

Phosphorus – is a substance that gives off light when it is electrically charged.

Lumen – the amount of light striking 1 square foot of surface at a distance of 1 foot.

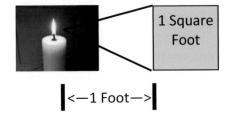

LED (Light Emitting Diode)

Background

LEDs have been used for lights on panels, video players, and other electronic devices as indicating lights. They come in many different colors. LEDs can also be used to illuminate spaces. LEDs used for area lighting are typically bright-white and are combined with a number of LEDs, called an array. Due to their long life and low power consumption LEDs are being used in new applications. LEDs are now commonly seen as automobile taillights, streetlights, and exit signs. They are now being used for task and spot or directional lighting applications. LEDs use the lowest amount of energy for the numbers of lumens produced.

Systems

Application Lighting

LEDs are great for certain applications where standard bulbs are subject to harsh environments and excessive vibration. Automobile taillights are a good example. Filaments of standard bulbs tend to break. LEDs do not have a filament and are long lasting.

Task and Spot Lighting

Replacement bulbs for low lumen or wattage applications have been developed. Many of the bulbs simply screw or plug into existing bulb sockets. Flashlights that use LED's can be used continuously for long periods of time before fresh batteries are needed.

Why is it Important?

LEDs use approximately one 10th of the electricity of incandescent bulbs. A typical 25-watt incandescent bulb gives off approximately 200 lumens. If this bulb were compared with a fluorescent of the same lumens, the fluorescent would use only 5 watts. For the same lumens, an LED bulb will use 3.5 watts. If all incandescent bulbs were replaced with low-wattage bulbs, a tremendous reduction in electrical consumption would be realized. Another factor is the life expectancy of each type of bulb. Incandescent bulbs will last about 2,000 hours and fluorescents will last up to 10,000 hours. LEDs will operate continuously for more than 100,000 hours, or more than 10 years. The use of LED lighting reduces the impact on the environment and energy resources.

Terminology

LED – a light emitting diode that emits light when connected to a direct current source.

Plumbing

Plumbing systems involve the distribution and use of water, a natural resource. Plumbing systems convey this very valuable resource to our homes and businesses. In many parts of the world, drinking water is a precious commodity and cannot be squandered. Water resources in many parts of America are taken for granted. In many places, where water is plentiful, these same water resources may become strained as our ever-growing population increases.

Reducing the amount of potable water used through innovation aimed at conservation, as well as rainwater capture, and grey water reuse has become more and more critical.

Moving potable water from the source to the point at which it will be used requires energy. The greater the distance potable water is moved the higher the energy demand. Electrical power plants usually provide this energy. Many of these power plants burn fossil fuels to create electricity. Large cities often pump water great distances to meet the needs of their inhabitants.

Green Plumbing systems not only conserve the precious resource of water, but also reduce the power requirements for moving that water, thereby decreasing our carbon footprint.

Grey Water

Background

Grey Water is not fit for human consumption that has not been used to move human waste. Water that is used for washing, bathing or laundry is grey water. Rather than sending the water down the drain, grey water can be used to water trees and plants. Grey water often contains heat. This heat can be extracted to be used to pre-heat cold water. In addition, grey water may be used to conserve potable water in applications where grey water would be acceptable. Example: Grey water could be used in commode tanks.

Systems

Grey Water Irrigation

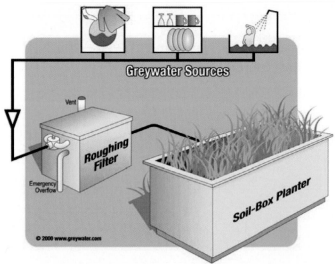

Grey water used for irrigation seems simple. However, it may be a little more involved than simply dumping used water on plants. Most grey water systems have holding tanks, valves, filters, and some may have automatic controls. The system must be able to store water for later use. If grey water was allowed to run freely, it might create puddles, ponds, and may run off the property in a stream. Storing and measuring the amount the water being used is one of the functions of a good grey water system.

Some pathogens (agents that may cause health issues) that exist in grey water may be transferred to some plants. Grey water is not recommended for plants and fruits that may be used for human consumption.

Grey Water Heat Recovery

Heat contained in hot grey water can be recovered from some grey water systems. Water heaters are used to boost the temperature of water to use for bathing. Bath water goes down the drain to a heat exchanger, which moves the heat from the grey water to the new cold water entering the water heater reducing the energy requirements for heating the water.

Why is it Important?
It is estimated that about 60% of water used becomes gray water. The water that goes down the drain could be used for other purposes. For example, the use of fresh drinking water for watering plants is wasteful.
Using grey water has the potential to:

- Reduce potable water consumption.
- Reduce the impact on septic and sewer systems.
- Reduce the energy used to pump and process water.
- Recharge the ground water.
- Increase the nutrient value of top soil.
- Stimulate plant growth.

There are also some concerns about the use of grey water. Pathogens that could be harmful to humans exist in some grey water. Some uses of grey water are better than others. Grey water can be used for flushing toilets, but the water must be first treated. Untreated grey water standing in a toilet bowl for long periods of time may create odors and present human health problems.

Terminology
Grey Water –is water that has been used for another process and not water that has been used to move human waste. It is water that is not fit for human consumption.

Pathogen – is an agent (bacterial, biological, etc.) that may affect human health.

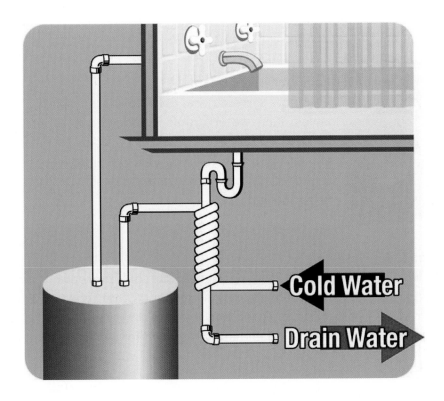

Rain Water Recovery

Background

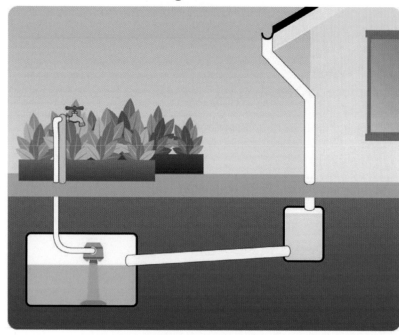

Rain water is a renewable resource that can be harvested, saved, and used for many purposes in place of fresh water. Rain water can be collected from rooftops, parking surfaces, and other large surfaces. Collected rain water can be stored in large tanks. Water from the storage tank can be pumped for:

• Watering plants and trees
• Watering lawns
• Washing cars
• Flushing toilets
• Commercial cooling tower make-up water
• Any other non-potable (nondrinking) water uses

In cities, rain water is sent to a storm drain. Rain water storage tanks are connected at the storm drain connection. When the tank is full the overflow is sent to the storm drain. Before reuse, collected rain water may need to be filtered to remove leaves, dirt, and other materials picked up during collection.

Systems

Rain Water Irrigation
Irrigation systems are used to water plants, trees and lawns. Rain water runoff is collected in large tanks and used to irrigate during dry times. Many of these systems are set up to water through an soaker-style hose. This system does not evaporate water as rapidly as the spray type irrigation system.

Rain Water Toilet Flushing
Toilet flushing systems can be a little more complicated. If the water is not treated it could begin to smell and create staining. Filters and other types of water treatment can reduce or eliminate the problem. If the rain water storage tank is positioned higher than the toilet, it will eliminate the need for a water pump.

Why is it Important?

The capture and use of rain water reduces the dependence on fresh water supplies. Reducing fresh water demands reduces the impact on sources of fresh water.

Terminology

Potable Water – is water used for drinking.

Storm Drain – is a collection point for water runoff.

Irrigation – is the process of adding water to the soil.

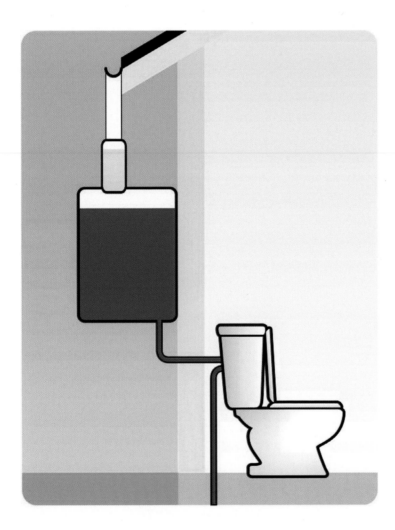

Potable Water

Potable water systems are plumbing systems that supply drinking water. Most plumbing systems supply potable water to plumbing fixtures, dishwashers, laundry equipment and showers. This section of the book will identify systems that heat fresh water for domestic use (DHW).

Instantaneous Hot Water

Background

Instantaneous hot water systems work on the principle of heating water on demand. These systems have no tanks to store water. They are also referred to as tankless water heaters. If sized properly they have the ability to work continuously, ensuring a constant supply of hot water. Tankless hot water systems do not have stand-by heat loss. Whereas tank-type water heating systems lose heat through the jacket during standby.

Systems

Gas

Instantaneous gas hot water systems are available in residential and commercial models. Gas tankless systems need a venting system for removal of combustion gases. For venting purposes, mounting the system on an outside wall is the most popular.

Electric

Both commercial and residential instantaneous hot water systems are available. Electric systems may be installed in any convenient location.

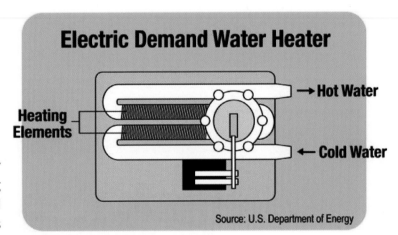

Steam

Another instantaneous hot water system uses steam. Steam systems are very efficient and are capable of rapidly heating large quantities of water. Commercial buildings that have steam heating systems are candidates for this type of tankless water heating.

Why is it Important?

Heating water on demand requires considerably less energy than that required by tank type heaters with stand-by losses. During times when there is low or no hot water demand, tank type water heaters experience heat loss through the insulating jacket of the tank. Most residential water heaters expend energy to heat and maintain water temperature when the house is empty. At times when there is a large demand, tank-type systems may not keep up with the demand.

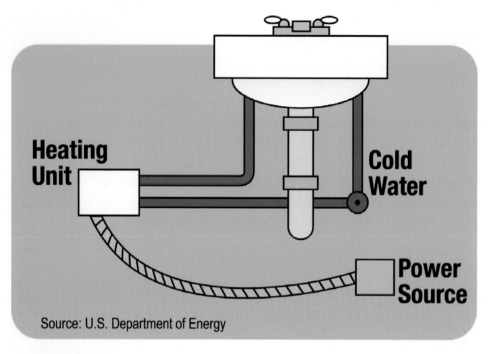

Source: U.S. Department of Energy

Terminology

Instantaneous Water Heater – also called tankless, are water heaters that heat water on demand.

Products of Combustion – are flue gasses, which result when a fuel is burned.

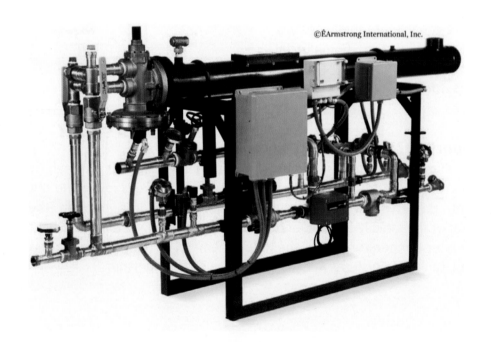

©ÊArmstrong International, Inc.

Solar Domestic Hot Water (DHW)

Background

The work of heating water can be done completely by solar means at various times of the year. At times when all of the hot water demand cannot be met, by solar heating, the solar water heating system can be used to preheat water. The solar water system preheats the water reducing the energy required by the standard water heater.

Solar water heating systems can be very simple. Hot tea can be made by setting a container of water and tea in direct sunlight for a few hours during the day. This is how some simple solar water systems work. The tank is exposed during the daylight hours and the sun heats the tank.

Many solar water systems are much more complicated. These systems must be able to produce hot water in colder climates and be protected from freezing. These systems use solar panels that are filled with a non-freezing fluid (typically propylene glycol, RV antifreeze). Pipes are connected from the panel to a heat exchanger that forms a barrier between the fluid in the panel and the water in the water tank. A small pump moves water through the heat exchanger and solar panel. The pump operates only during times of the day when solar energy can produce heat.

Systems

Direct Sunlight Tank Heating

There are better systems than the one depicted, but the illustration of making sun-tea shows how simple these types of solar water heating systems can be. Other types of systems enclose the tank during the night with insulating doors. All of these systems are manual. These systems are also called "batch" collectors because they heat all of the water in one place and in one container.

Thermal Siphon

Thermal siphon systems use the natural tendency for warm water to rise. Rising warm water is captured at the top of the system in a tank. Cooler water from the tank drops to the bottom of the solar panel to be heated by sunlight as it rises to the top of the panel. These systems work independently and do not require any other type of energy to pump water through the solar domestic hot water system.

Forced Circulation Systems

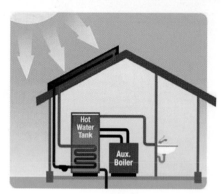

A forced circulation system uses a pump to move a freeze-protected fluid through the solar collector and to a heat exchanger. Another pump may be used to move water from a storage tank to the heat exchanger. These systems can be installed in many different arrangements. The solar panels are the only things that are visible and can be mounted on the roof or the ground. The pumps can also be operated by using solar produced electricity.

Why is it Important?

Solar water heating is one of the easiest and most efficient uses of solar energy. These systems are able to pay for themselves in a few years (depending on costs and usage) and have low maintenance requirements. Solar water heating system components can last for 20 years or more under normal operation. Solar water heating systems, when used in conjunction with traditional systems reduce the electrical and / or fossil fuel requirements for water heating.

Terminology

Solar Water Heater – is a system used to heat water with energy derived from solar radiation.

Thermo Siphon – the effect of water movement (water that rises when heated and drops when cooled).

Forced Circulation – means that a pump is being used to move the water in the system.

Batch Collector – is a type of collector that stores and heats a quantity of water.

Reduced Water Volume Systems

Reduced water volume systems run the gamut from low water usage to no water usage. Reducing the amount of water used for human waste disposal has the potential to save hundreds of gallons of fresh water, thereby reducing the energy required to move and purify the water. The following systems highlight some of the reduced water usage systems available on the market today.

Waterless Flush

Background

Waterless flush devices are being used extensively around the world and have been proven to work effectively. Waterless or no-flush urinals do not use water to flush away human waste. These devices are designed to allow urine to pass through a trap without using water to flush it down the drain. The trap is specially designed, to allow urine through but seals out sewer gas. In many cases, the trap and other parts can be removed from the urinal for maintenance. Regular maintenance requires a light spraying of cleaner and wiping each day.

Cross - Section of the Patented Vertical EcoTrap®

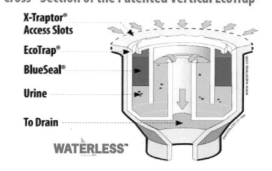

Systems

Cartridge Trap

This system uses a cartridge filled with fluid. The fluid allows urine to pass through, but seals out sewer gasses. The fluid requires periodic replacement or "topping-off."

Mechanical Trap

Some of these units use a mechanical trap that allows urine and other fluids to pass through, but seal out sewer gasses. Some units have a device that forms a vertical "S" trap. In both cases, the trap requires periodic cleansing. The trap is sometimes called an "insert" and trap parts can be replaced during major maintenance.

Why is it Important?

Each urinal can save as much as 45,000 gallons of water annually. This staging amount of fresh (potable) water savings can result in a tremendous energy reduction.

Terminology

Waterless Flush – is a device that does not use water.

Trap – is the part of the system that seals waste products and gasses from occupied spaces.

Low Flow

Background
Low flow plumbing devices can be attached to many types of fixtures. Some of the most popular are devices that replace faucet aerators and shower heads. These devices reduce the volume of water being used. Many of these devices are simple restriction devices or orifice plates. Water flow is restricted because of the size of the hole in the plate. Aerator devices mix air with the water, thereby increasing velocity, which gives the impression that the flow has not changed.

Systems
Showerheads
There are many different designs of low flow showerheads. Some of these maintain a high-pressure spray while others try to imitate rain. Spray types generally include concentrated, pulse, rotating, and gentle. Consumers have a number of choices to meet their personal preference. What is important is the amount of water flow per minute. A conventional showerhead uses 15-19 liters of water (3- 4 gallons) or more per minute. A low-flow showerhead uses approximately 8 - 9 liters of water (2 – 2.3 gallons) per minute.

Faucets
Low flow aerators can be attached to most faucets. The aerator mixes air with water to give the appearance that there is a large volume of water. Every aerator has a flow rate stamped on the side.

Why is it Important?
Reducing water consumption and the amount of energy needed to pump and condition the water is the objective of low flow devices. A typical 10-minute shower uses 42 gallons (190 Liters) of water. Low flow showerheads can dramatically reduce the amount of water used. Low flow water devices reduce the impact on water resources, and save energy.

Terminology
Low Flow – is a device that reduces the volume of water measured in gallons or Liters per minute.

Fixture – is one of many types of plumbing devices used for water distribution.

Aerator – is a device that mixes air with the water stream.

Low Flush

Background

Low flush fixtures reduce the amount of water used per flush. Although the public reaction concerning low flush technology has been mixed, low flush devices that are now in place have saved thousands of gallons of water per year. The object of these units is to complete the flushing and bowl washing operation with as little water as possible.

Systems

Pressurized

Many pressurized-low flow units use 6 liters (1.6 gallons) of water or less. Some pressurized tank units use water to build pressure against an air bladder in the tank. The air bladder is compressed when not in use. When the unit is flushed, the air expands and provides an extra force to complete the task of flushing and washing the bowl. Other pressurized units use a pump to pressurize or assist in the flushing operation.

Gravity

Low flush units use .8 to 1.6 gallons of water (3 to 6 Liters) per flush. Some are dual-flush units, meaning that they can use the minimum amount of water or a greater amount of water when necessary. These units rely on gravity to perform the flushing and washing operation.

Metered Flush

Metered flush units use a flow control rather than a tank. Tank-less units use the same type of flow control as some urinals. The flow control meters the amount of water usage by time. The valve is open for a period of time that may or may not be adjustable, depending on the valve and manufacturer. Metered flow units are typically used in commercial applications.

Why is it Important?

Using fresh (potable or drinking) water to move human waste is considered an inappropriate use of fresh water. It is the objective of many manufacturers of toilet fixtures to reduce the amount of fresh water used for this operation. Toilets account for more than 30% of indoor water usage in an average home. Estimates vary, but many indicate that 4,000 to 7,000 gallons of water could be saved by updating to a low flush fixture.

Terminology

Low Flush – is a plumbing fixture that uses .8 to 1.6 gallons (3 to 6 Liters) of water per flush.

Appendix

Biodegradable or Biodegradation

Time Required for Decomposition

Cotton Rags	1–5 months
Paper	2–5 months
Rope	3–14 months
Orange Peels	6 months
Wool Socks	1–5 years
Cigarette Butts	1–12 years
Plastic Coated Paper Milk Cartons	5 years
Plastic Bags	10–20 years
Leather Shoes	25–40 years
Nylon Fabric	30–40 years
Tin Cans	50–100 years
Aluminum Cans	80–100 years
Plastic 6-Pack Holder	450 years
Glass Bottles	1,000,000 years
Plastic Bottles	Forever

Background

Nearly everything degrades. The breakdown process depends on the amount of water and sunlight they receive. Water and sunlight help to stimulate the various biological processes in nature that helps break things down. The term "biodegrade" is an over-used term that needs some additional explanation. The standard for biodegradable was set by a European organization that developed several standards regarding products and their ability to biodegrade or decompose. The British Organization for Economic Cooperation and Development (OEDC) defined Biodegrade (or Biodegradation) as, "the process by which organic substances are decomposed by micro-organisms (mainly aerobic bacteria) into simpler substances, such as carbon dioxide, methane, water and ammonia". There is research being conducted today to determine how packaging products can be made to break down into these common substances. For instance, a few plastics are made from starch are considered biodegradable.

Why is it Important?

Nearly everything ends up in waste systems. Some materials are acceptable, while others can be tolerated; but many other substances and materials are not compatible with the waste system and do not break down or break down very slowly. New materials are continuously being developed and coming into the market. These new materials have labels to indicate their ability to biodegrade. Consumers need to make decisions when buying products and containers, about their biodegradable capabilities. Consumers have an effect on the products and packaging brought to the market place, via their buying power. Wise purchasing can mean a Greener environment.

Terminology

Biodegrade (Biodegradation) – Substances decomposing into simpler substances and disbursing into the environment as naturally occurring elements.

Elements – Common basic substances, such as oxygen, carbon, iron, etc.

Aerobic Bacteria – A bacteria that needs air to live.

Microorganisms – Very minute living things that are difficult to see without using a microscope.

Starch – A food substance created naturally in plants.

Post Consumer Waste

Background

Post Consumer Waste is the products left- over after a product is purchased. This includes packaging, such as bottles and boxes used to market and carry consumer products. It also includes discarded or used products, such as computers, bicycles, or clothing. The term most often used to describe post consumer waste is "garbage". This includes solids, liquids, and some containers of gases. Post consumer waste could be anything that has been discarded. Recycling is the logical solution to the ever-increasing amount of waste. Reclaiming is another way of reducing certain types of waste. The Environmental Protection Agency (EPA) has twelve tips for reducing solid waste.

REDUCE
1. Reduce the amount of unnecessary packaging.
2. Adopt practices that reduce waste toxicity.

REUSE
3. Consider reusable products.
4. Maintain and repair durable products.
5. Reuse bags, containers, and other items.
6. Borrow, rent, or share items used infrequently.
7. Sell or donate goods instead of throwing them out.

RECYCLE
8. Choose recyclable products and containers.
9. Select products made from recycled materials.
10. Compost yard trimmings and food scraps.

RESPOND
11. Educate others on source reduction and recycling practices.
12. Be creative - Find new ways to reduce waste quantity and toxicity.

Why is it Important?

The disposal of post consumer waste requires energy, land, and other resources. Therefore, reducing this waste, through smarter packaging, reuse and recycling reduces energy consumption and our carbon footprint.

Terminology

Post Consumer Waste – is waste produced by the consumer or end-user of a product. It may include packaging, portions of the product as in food waste or the whole product upon obsolescence.

Reclamation

Background

From 1901 to the 1970's, the term reclamation was used to mean "the damming of rivers to control water". Today, reclamation is used when talking about waste water. Waste water can be restored or used for useful purposes (such as watering plants and trees). This would be considered reclaimed water. However, the term reclamation can be used in conjunction with anything that is restored to a useful condition. The Bureau of Reclamation is concerned today about the sound use of water resources, and is sometimes referred to as water conservation.

Land is also reclaimed. Land reclamation is the restoration of land from mining or industrial operations that have left the land in an unusable condition. Open pit mining and other industrial operations have left the land in a "brown field" condition. These are only two examples of land needing restoration.

Why is it Important?

Resources that can be reclaimed, almost amazingly become useful once again. Of course, it is more important to understand that contamination can occur through negligence. The dumping of oils and other forms of contamination on the ground adds to "brown field" conditions. This type of contamination requires a considerable amount of energy and recourses to reclaim.

Terminology

Reclamation – The process of bringing something not useful, back to a useful condition.

Brown Field – Contaminated land or soil which cannot be used until reclaimed.

Solid Waste Management

Background

Solid waste is any waste that is not a liquid or a vapor. The vast majority of these are food, metal, glass, rubber, plastic, and paper products. And among these, are hazardous waste products which pose health risks to living organisms. Municipal (city) waste is sometimes referred to as MSW (Municipal Solid Waste). Municipal waste is typically garbage. In 2005, the EPA reported a total of 245 million tons of garbage. This is about 4.5 pounds of waste per person per day. Cities spend a large portion of their budget on municipal waste. Residential waste is a part of every city's disposal program. Households throw away common items such as paints, cleaners, oils, batteries and pesticides. These items contain hazardous components and can cause sickness, if released to the environment. (Also see Post Consumer Waste.)

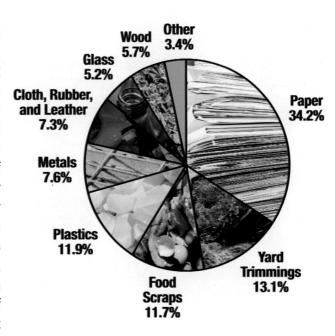

The Resource Conservation and Recovery Act (RCRA) was enacted by Congress in 1976 and amended in 1984. The act's primary goal is to protect human health and the environment from the potential hazards of waste disposal. The RCRA calls for the conservation of energy and natural resources, reduction in waste generated, and using environmentally sound waste management practices (EPA, 2007).

Why is it Important?

Reduce, reuse and recycle have become the "3R's" to managing waste, according to the EPA. To be aware and to try to reduce the amount of solid waste before it becomes part of the waste stream, seems to be the first step. It is important to be mindful of the risks to the environment and human health that certain chemicals and ordinary items possess.

Terminology

Solid Waste Management – An effort to control and move solid waste (garbage) from the point of disposal to an area of control.

Hazardous waste – Any waste product (solid, liquid, or gas) that has the potential to cause illness to any living organism.

Organism – Life forms that depends on the environment for food and shelter; the life form could be anything from a single cell to humans.

MSW – Municipal Solid Waste.

Sustainable

Background

The term sustainable (or sustainability) means that the sustainable process whatever it is will be able to be repeated continually into the future. In the purest sense, the process would have no impact on the environment and other resources to be completely sustainable. Brundtland Commission defined sustainability is what "meets the needs of the present without compromising the ability of future generations to meet their own needs". Other ways to use this word are in reference to being able to create a repeatable process. Creating a sustainable process would mean that other parts of the process would also need to be repeatable. Each part would not reduce or eliminate any resource. Sustainable has links to food, housing, energy, and environment.

Why is it Important?

Where ever possible, sustainable processes should be chosen over those that would deplete resources. The well being of the next generation is important to us all.

Terminology

Sustainable (or sustainability) – repeatable without effect or impact on current or future resources.

Sustainable Process – is an activity which involves food, shelter, comfort, transportation, etc. to meet human needs.

Volatile Organic Compounds

Background

Volatile Organic Compounds (VOC's) are part of many mass-produced items for consumer use. VOC's are chemical compounds that evaporate (outgas) or emit from commonly used materials, such as cleaning agents used in businesses and homes. These organic gasses can be emitted into the air and react with other pollutants. Along with particulate matter, sunlight and other pollutants can work together to form ground-level ozone; a main ingredient in smog. These gasses can cause many human health issues, depending on the levels or amounts of gas (or gasses) present. Some of these gasses may come from paint thinners, cleaning solvents, and fuels. Others come from

natural sources such as trees, animals, and insects. Of these, the EPA (Environmental Protection Agency) lists the following as major sources of concern: paints, paint strippers, and other solvents; wood preservatives; aerosol sprays; cleansers and disinfectants; moth repellents and air fresheners; stored fuels and automotive products; hobby supplies; dry-cleaned clothing.

Why is it Important?

Reducing the amount of air pollutants is good for human health. When using products that pollute the air within a building, the products often become trapped and become a breathing hazard. Choosing chemical agents, materials, and products that have low or no VOC's is important to any building or home environment.

Terminology

VOC (Volatile Organic Compound) – A compound that can send out-gas to the air in our environment. These compounds are or can become air pollutants.

EPA (Environmental Protection Agency) – A governmental agency charged with a mission to protect human health and the environment.

Compound – A molecule that contains at least two different elements; hydrogen; (H2) and oxygen (O) form the compound H2 O, water.

Particulate Matter – Particles of substances that can be carried by air currents, but if the air is still, can drop out of the air like dust.

Out-gas – is the gas that leaves a material as it ages or cures.